MORE JOURNEYS BENEATH THE EARTH

The autobiography of a cave explorer.

Volume 2.

David William Gill

Copyright © 2021: David William Gill

All rights reserved.

ISBN:

I was also a caver during the years of Dave's caving all over the world. I was living my dream as he was living his dream life of cave exploration. As good as mine has been, I would gladly have traded lives with him for the amazing things he's done.
C. William "Bill" Steele, Texas USA

The first time I met Dave Gill was in Papua New-Guinea, in the Nakanai Mountains of New Britain to be precise, at the beginning of 1985. I was on my way back from the Minye mega-doline and stopped off in the village of Nutuve where the British team led by Dave Gill had established its camp. Dave had managed to set up a team to tackle the Nare giant cave and its roaring underground river but also newfound caves. I expected to meet a tall man so to my surprise, Dave is rather a short man, but I knew he was a great caver, one of the best British explorers who had been caving in many karstic areas around the world.

So, it was not a surprise that our second meeting took place in the heart of Borneo pristine forest where Dave was in charge of the Mulu caves national park, an area famous for its giant underground networks. In 2006, Dave managed to find sponsors for another expedition to the Nakanai Mountains. I was delighted to be chosen by Dave to be part of the team. Under his leadership, the expedition was greatly successful. The Ora-Phantom pot and Mageni caves explored during the trip stand out among the best Nakanai caves, and consequently, the BBC charged Dave to form a team in 2008 for the caving part of the "Lost Land of the Volcano" documentary. The shooting took place in Mageni cave.

To have been able to share these adventures with Dave has been a fantastic experience and an honour for me.

Old age might reduce the chance for us to meet again on the ground but a project like the "Sublime karst of Papua New Guinea" in which Dave has his voice could be an occasion to meet again and to succeed in the protection of a karstic area we both love, the Nakanai Mountains.
Jean-Paul Sounier, France.

Few explorers in the world today have set foot in as much undiscovered terrain as veteran caver Dave Gill, who has opened up sights of grandeur and beauty from the Derbyshire Peak District to the remote jungles of New Britain.

In this, the second enthralling and lavishly-illustrated volume of his memoirs, he reveals further wonders, focusing especially on exploration of the vast and magnificent caves of the place that has become his second home – the rainforest massifs around Mulu in Sarawak. He also tells the story of how he has put something very valuable back, helping to run the Gunung Mulu national park for many years and successfully drawing up the plans which have led to its significant northern extension, Gunung Buda. These great reserves make up one of the world's finest tracts of tropical forest, teeming with every kind of wildlife, and their preservation in a pristine state for future generations is another of Dave's achievements.
David Rose, Oxford, UK.

Dave and I started our Mulu 'journey' back in 1988 when the Gunung Mulu National Park was in its infancy, a very different place, largely unexplored and remote from civilisation. We naively thought it would be a one off experience; how wrong we were.

Dave's recollections bring back the magic of the place, the people, the comradeship, the discoveries, the near misses, the friendships made and the hardships endured to explore massive caves in a tropical climate. His is an extraordinary story, dedicated not only to the world of cave exploration but latterly to the people of Sarawak who featured so prominently in his life.

Together we had the fortune to participate in the exploration of what proved to be some of the most fantastic caves in the world. Our lives would have been very different had Mulu not brought us together all those years ago.
Matt Kirby, UK

There are special people in this world whose life experiences are beacons of encouragement to others, no matter what the area of expertise. All dedicated cave explorers will tell you their pursuit is 'a way of life', but few individuals are privileged to put that literally into practice. Dave Gill is one of these: his accounts of expeditions all over the world, often in circumstances of incredible danger, provide the reader with a perfect narrative of excitement, science and technical expertise. The caves described within this volume should be regarded as cathedrals of nature –

places of consummate beauty, housing bizarre, fragile rock formations unique to the planet, perhaps the last truly unspoilt terrain on earth and, being situated in largely impenetrable rainforest, hopefully to remain so.

However, this book offers much more than an underground travelogue. Dave is a passionate man, his assessment of how mankind is desecrating our planet burns brightly from every page. It forms yet another desperate *cri de coeur* to halt the spread of desecration needlessly threatening our world in the name of profit. He has worked tirelessly against bureaucracy for over forty years to set up National Parks and conservation schemes that will preserve these wonderful caves and their surroundings for the future. It has been his life's work. Hopefully this book will serve to guide future adventure seekers along similar paths before destruction becomes irreversible. As a bonus, it takes the reader into places as alien as Mars, with a crowd of excellent photographs to illustrate a narrative where it would be so easy to lapse into hyperbole yet presents a balanced account of what puts the fire into every caver's soul. It should be required reading at government level, especially in those countries fortunate enough to possess such stunning cave systems.

Alan L. Jeffreys, Grampian Speleological Group, Edinburgh, Scotland.

Dedication

To the many great cave explorers who are no longer with us. I was fortunate enough to know them and explore great and outstanding caves with them over the past 50 years.

The author mapping in the giant cave of Beimo Dong China.
Photo: Jerry Wooldridge

Contents

	Acknowledgments	ix
	Introduction	1
1	10 year in the Borneo Jungle	3
2	Karst Management Unit, Sarawak	20
3	Caves and karst regions of Sarawak	29
4	Gunung Api Connection Expeditions, 1990 to 1994	47
5	Reconnaissance to Hidden Valley 1993	52
6	Hidden Valley expedition 1996	56
7	Hidden Valley 1998 expedition	65
8	The Japanese Expedition 1997	70
9	Gunung Buda, forming a new national park	74
10	USA Gunung Buda Expedition 1995	81
11	Gunung Buda USA Expedition 1996	92
12	The Gunung Buda Caves Project 1997	99
13	Gunung Mulu World Heritage 1999	113
14	USA Gunung Buda Project 2000	119
15	Goodbye Sarawak	128
16	French World Heritage Caves Nomination	132
17	A few cave and karst projects	139

18	Untamed Rivers of New Britain 2006	145
19	BBC filming of Mageni Cave, New Britain 2008	172
20	Back to Sarawak	181
	Epilogue	187
	About the author	188

Acknowledgments

As with the first book I am extremely grateful to the photographers for allowing their photographs to be reproduced in this volume. Dave Bunnell, Peter Bosted, Matt Kirby, Robbie Shone, Andy Eavis, David Nixon, John Gunn, Rambli Ahmad, Liang Uvang and special thanks to Jerry Woodridge for enhancing all the photographs and designing the cover. Appreciation must go to Brian Thomas, Alan Jeffreys and Matt Kirby for their splendid proof reading skills. Finally and by no means least my thanks to Sarawak and its people for the many exciting years and giving me the unique opportunity to contribute.

Cover photograph by Jerry Wooldridge: Firecracker passage, Blackrock Cave, Gunung Mulu National Park.

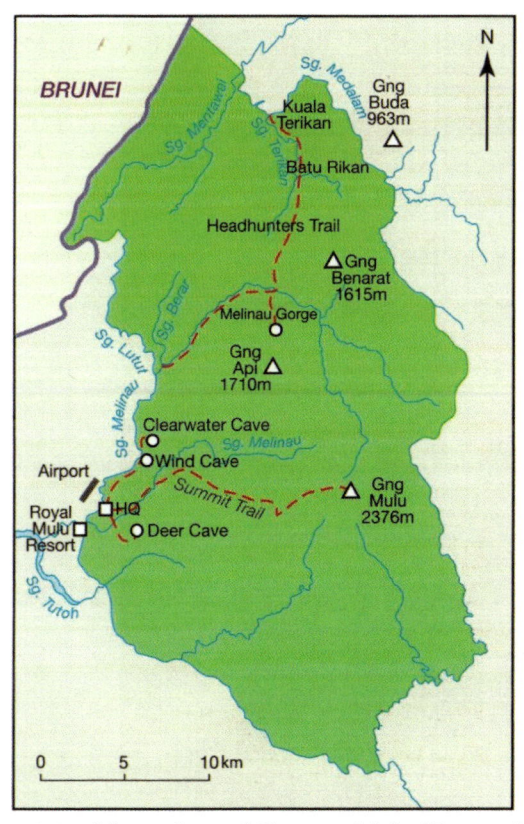

The original boundary of Gunung Mulu National Park

Introduction

Over 50 years of exploring caves from poking my head into a cave in Derbyshire with a bicycle lamp and no helmet to leading expeditions to the largest and most dangerous caves in world, yes caving has become a lifelong obsession. The first volume created some interest so I thought; as I had run out of work in Sarawak I might as well write volume 2, the second 25 years. So here it is.

The first part of this story 'Journeys Beneath the Earth' ended on my departure from the UK at the age of 50 in order to take up an entirely new profession. The appointment as a speleologist and development officer enhanced my lifelong love for Sarawak and especially the spectacular Gunung Mulu National Park. With its primary rainforest, biota and its huge complex cave systems, Gunung Mulu can only be described as an outstanding example of tropical karst terrain. The following stories relate to the work and expeditions during this ten year period and the following years back in England where I just had to get back to New Britain, Papua New Guinea. After ten years in the UK I returned to Sarawak and I am now presently privileged to work on a voluntary basis on conservation, national parks, world heritage values and projects for the local communities.

I have also been lucky enough to have travelled the globe and work in exotic places and have seen first-hand the devastating tragic destruction of the planet and the unsustainable development coupled with an alarming unsustainable rise in population figures. We live on a beautiful planet with a huge diversity of animals and plants and wonderful natural wonders. Being involved in conservation and attempting to save what little high biodiversity areas we have left, I have come into contact with many fine men and women working in the field of conservation. All agree that we are flogging a dead horse but to sit back and do nothing is a crime in itself, it is better to try and fail than not try at all as our children's children will ask why we sat on our back sides and did nothing? The demands on land and water supplies coupled with financial greed, unsustainable hunting, sale and consumption of wild animals, unsustainable fishing and destruction of coral reefs, destruction of rainforests, global warming, pollution and the continuing loss of species can only lead to eventual catastrophe for the planet and inevitably the human race. Many conservationists agree that we are now over the brink and on the downward spiral to inevitable catastrophe which is a dismal prediction.

Outer space and the deep blue oceans are all the preserves of highly technical and expensive exploration, but cave exploring and conservation can be accomplished on a shoe string by ordinary people.

Her Royal Highness Princess of Thailand Maha Chakri Sirindhorn, known in Thailand as Phra Thep, Princess Angel, a conservationist with the author, Gunung Mulu National Park, Sarawak.
Photo: Laing Uvang.

1

10 years in the Borneo rainforest

'Choose a job you love, and you will never have to work a day in your life.'

Confucius (551 BC-479 BC)

In 1991 I decided before leaving the UK to throw a party that the Eldon Pothole Club affectionately calls a 'Stomp'. For those that have no idea what this means it's a dance with a difference. It consists of a very loud rock group with many cavers dancing around half naked with much guzzling of beer. In this case it was held at the Bull i'th Thorn public house on the Ashbourne road Derbyshire with our favourite rock band, 'Eric and the Frantics' who were without doubt extremely frantic. The Eldon Pothole Club had been banned over the years from having stomps and Christmas Dinners in most of the public houses in Buxton, but the Bull i'th Thorn seemed to put up with the mayhem as they sold lots of beer. A local newspaper advertised the stomp so the venue was packed with caving friends from around the country. The usual proceeding was to strip naked the party giver and throw beer over him. It was a great send off with a mammoth headache the next morning.

In 1991 Gunung Mulu National Park came under the jurisdiction of the National Parks and Wildlife Department, a section of the Forest Department Sarawak with the Head Office in Kuching. Under the State Government law the Director of Forests has jurisdiction over all Protected Forest Estates and Totally Protected Areas under the relevant ordinances published in the Government Gazette. First thing the chief park warden instructed me to do was to sort out the electrical installations in the four show caves as nothing worked. He was not joking as it was a mess. We had a sub office in Miri where the park warden for the northern parks had his base. So I worked under the direction of Oswald Braken Tisen, an energetic and keen individual with a passionate interest in conservation and he also liked caves! Things today have changed as national parks come under the jurisdiction of the Sarawak Forestry Corporation.

Gunung Mulu National Park was still in the development stage in 1991 and things were pretty basic. We had no airport or roads so the only way to get there was a one hour taxi ride from the town of Miri to Kuala Baram where a 7 am ferry boat could be caught with a two and a half hour boat trip upstream on the Baram River to Marudi town. A stop over for a few hours was required before the ferry boat left for a three hour journey to the Longhouse of Long Terawan situated on the Tutoh River, a major tributary to the Baram River. With luck a longboat could then be hired for a one and a half hour journey upstream to the park headquarters situated on the Melinau River. It was a long full day journey but fun as the river was of interest to visitors.

The park had no telephone or internet as it had not yet been invented so all communication was via the postal service. The closest post office was a three day boat journey there and back. Electricity was via diesel generators which were only operational during the day, 11 pm was switch off time. We had a single side band radio transmitter with a difficult connection to our head office in Miri and a VHF transmitter on the summit of Gunung Mulu. This worked locally around some areas of the park and was powered via a rechargeable lead acid battery. Two maintenance staff had to climb the mountain occasionally to charge the battery, a four or five day journey! There was no such thing as shops where you could purchase food supplies, just a few cafes mainly involved in selling beer. If you required money from the bank or food and drink, it was a three day trip to Marudi and back.

The main problem I was going to face was the fact that I was not in charge of anything. The park was managed by the Officer in Charge who worked under his superior based in Miri. I envisaged that it was going to be difficult to achieve anything without support which proved to be correct.

I owned an old Amstrad computer, in those days very modern with 72 kilo bits of memory and also an ink printing machine which I managed to drag all the way to Gunung Mulu. At least I could do my own typing with modern technology so I set up my office in the administration building. The Officer in Charge was at that time Victor Luna Anam. I had met Victor during the last expedition and we had a good working relationship and he also liked caving which was a plus. I stayed at Victor's house for a few days until I was given my own quarters next door to our boatman, Siduk. To make matters more interesting Betty lived next door, the lady I had met previously who had just started working at the park as a housekeeper.

I started work immediately inspecting the four show caves and generators. The Clearwater lighting installation had completely burnt to a frazzle, impossible to repair and the floating bridge across the Clearwater Cave River was in danger of being washed away during heavy rain. Nothing worked in Deer Cave and the three cave generators were antique. The Deer Cave lights were aluminium spot lights corroded via all the bat guano and urine continually dropping from the roof of the cave and the very smelly plank walk ran beneath the bat colonies so became inches thick with bat droppings. The control panels in the caves were composed of steel highly corroded boxes with poor control and wiring. The contactors in the panels continually rattled. Wind Cave required a complete re wire as did Langs Cave. After a long inspection of the show caves I submitted my reports and recommendation but no action was taken. Eventually I was instructed to attempt to fix it all myself. It was obvious that the existing electrical installations could hardly be called professional. The problem with the tropics is the high humidity which includes the caves, so all electrical apparatus is wet and nothing installed was damp proof. Touching any control panels in the caves and you received an electric shock. It was going to be a long job but after many months I managed to get some form of operation going with upgrading and water proofing the electrical installations with replacement of lights and panel controls. Being the only engineer at the park I had to do this myself with the help of one of the maintenance staff. Working in the caves with carbide lighting was not easy.

One of the major problems associated with show cave development is the essentiality to use sustainable materials as caves should only be developed once due to damage during construction. The walk ways were composed of timber which rots and produced fungi and bacteria detrimental to the cave environment. Continuous maintenance was required repairing broken wooden planks. Not ideal but there was little I could do on upgrading the walk ways without major reconstruction; all we could do was continuous repairs.

As for the staff it was inadequate considering the 7,900 visitors recorded in 1991. We required many more maintenance staff and housekeepers. The ten guides were local Berawan and good at their job but were not employed directly by the park. They were freelance and paid at a standard rate by the visitors or tour operators; after some enquiry I found out that they were owed thousands of ringgit from the tour operators. This had to change.

The water supplies were also a mess with old square galvanised metal containers which of course leaked. The obvious choice is plastic circular water containers so the weight of water is distributed. The water was pumped from the river to the tanks without any form of filtration. Most of the staff collected rain water via the roofing of their quarters. I also searched for fire hydrants but could not find any.

The diesel fuel was transported all the way by boat from Marudi town in steel drums which had to be hauled up the river bank and rolled along the path to the generator house, a huge and long task for the maintenance staff. In the dry season we ran out of fuel as the boats could not proceed upstream passing the rapids in the Tutoh River. No fuel meant zero electricity. The list of problems to solve was becoming longer every day. I calculated that I spent four days per week maintaining the park as there were no engineers stationed there. The list of problems that required solving ran to a couple of pages which I submitted to Miri office and Ministry of Tourism. It seemed as if I was becoming a maintenance engineer!

The total number of staff was around 42 people, all from the Berawan community but I was lucky to have some good men. The late Tama Bulan, a well-known cave explorer in his day, had worked with the previous expeditions and he was the maintenance supervisor. He proved to be a great help. Siduk was our boat man, again highly reliable with helping me to sort out the numerous problems. Franklin was the store keeper and often came with me to maintain the park infrastructure.

The men in charge were Forest Guards; there were no Park Rangers so their knowledge of operating a complex system with caves, pathways, accommodation and many visitors was limited. We had three stationed in Mulu, one in charge of stores, one accommodation and the other maintenance. Sadly they were seen in the morning and for the rest of the day quite often disappeared. The problem was that the Forest Guards had been trained in forestry which is where they wanted to be, not permanently stationed at some park isolated in the jungle and dealing with tourists. Visitor and park management is a totally different profession. I presumed that much of my work relied on not just supervising but actual hands- on work. The Minister of Environment and Tourism finally made an appointment to help me with finances in order to upgrade the infrastructure. Garry Tay was a huge help in obtaining the permission and budgets for planned up-grades.

I set up a small clinic next to the store as anyone feeling sick needed to travel for a few days to Marudi hospital. This worked well and became popular as we had a very competent medical orderly.

When we managed to get all the lighting working in Wind Cave the generator became overloaded and cut out. No one had calculated the total load-which was 15 kilowatts-but the generator was only 15 Horse Power. I obtained a new 20 kilowatt Cummings generator and installed it with great difficulty as getting it onto site was a problem; this ran all the lighting well. I did the same with Deer Cave and the maintenance staff spent many hard days rolling this very heavy generator all the way to the cave using wooden rollers.

New accommodation along with a canteen was under construction so, after completion, I inspected every part carefully and reported faults which were dealt with adequately. We therefore opened the accommodation for visitors and the canteen under a private concession agreement.

Due to the new accommodation the park generator was inadequate to run all the electrical installations as the VIP chalets had air conditioning. It was decided to run two generators in synchronous phase without realising that any change in the phase of one generator creates serious imbalance. This means that a trained operator needs to be on station permanently for any adjustments. As we had no engineer who could accomplish this task it proved impossible. Once more I arranged for the purchase of a new 120 kilowatt generator which was sufficient to run the park headquarters.

Adventure caving was one of my specialities so, with park guides, we explored numerous caves to set up tours. I imposed training of the guides with maximum tourist numbers at six and it proved to be popular. We even allowed tourist trips into Sarawak Chamber-the largest underground chamber in the world-depending on weather conditions as the underground river can flood during heavy rain.

The police decided that two men should be stationed at the park so we converted one of the quarters into a police station but it seemed to me that it was used as a casino. One of our Forest Guards was a cock fighting fan and had six very noisy ones so cock fighting became a Sunday pastime. I managed to get it stopped at the park so the fans of cock fighting moved to the other side of the river.

It was not all work though as we had plenty of fun. Sadly Victor left after six months to study for a degree. By this time I had inspected and opened to the public all the new accommodation which included the privately-operated canteen where we spent many happy nights drinking heavily and singing to the Karaoke. We also had a drinking spot downstream where we spent our Saturday nights. There were no lights so we used cigarette lighters in order to navigate the river-which was madness on the way back with a boat load of drunken park guides. We did not have a bridge across the Melinau River so, to get to the bar on the true right bank which was outside the park, we shouted for a boat to cross and pick us up.

Whenever we had a birthday party it was held beneath the quarters and usually turned out as a drunken affair often ending in a fight. In these early days Mulu can only be described as a little like the Wild West.

Life was anything but boring as my friendly relationship with the park staff grew exponentially as well as my friendship with Betty. We were married within the year and set up house together. The marriage party was organised by Betty and her relatives and was held in Miri with a few hundred guests. The general cultural activity of the Berawan ladies is to get all the men drunk which they succeeded in doing admirably. A kind friend from Tropical Adventure supplied thirty crates of beer and six bottles of whisky. We also had a few gallons of the local brew known as Tuak, which is a rice knock out wine which everyone is forced to drink. The party was a success as those collapsing were carried out to cars along with a helper to see them home.

Betty's late father was a famous Berawan resident called Usang Ipoi who was responsible for building the Royal Geographical Society base camp for the 1974/75 expedition. He gave us a piece of land which he farmed for rice. This had been in the family for 100 years or so and had a gigantic isolated cliff of limestone and a cave called Gua Usang. A cave in our garden sounded good to me so over the next five years we built a large house where he could live.

The culture of different communities was a subject that interested me having taken part in and led many expeditions to remote areas but the understanding of the communities of Sarawak was an uphill struggle. It takes years living with the people to comprehend how different the culture is and coming to terms with it. The elders within the community still believed in the spirits of the jungle and one day they made a ceremony across the river opposite the park headquarters. They summoned eagles to ascertain the traditional rites of the community land depending on the direction of flight of this magnificent bird. The eagles came and flew off in the correct direction. Betty had a grandfather who had fought in the war against the Japanese. The Englishman Tom Harrison-the leader-gave them permission to once more collect heads. Grandfather had a head hanging up in the Longhouse and the parang he had used still with the human hair wrapped around the handle. The children all swore the parang sang during the ceremonies conducted by the elders. After grandfather passed away in Miri City we loaded his coffin onto the back of a truck to take to the grave yard. I looked up and saw a flight of eagles hovering not far away. This seemed strange in the middle of the town. They have just come to say goodbye I was told, they came back the next day but when told that grandfather was dead they flew away. Sadly the head and the magic parang were disposed of after his death.

Another story was recounted to me by Mike Meredith the development officer before I took over. They were busy installing the plank walk in Wind Cave but every time they took the small generator into the cave it would not work. Outside the cave it worked first time. It was suggested that the cave spirits were at work and a ceremony was required including a sacrifice of a chicken at the cave entrance. The sacrifice was conducted and from then on the generator fired up first time inside the cave.

An interesting story revolves around Betty's mother and father. They were both born on the same day, 31st December 1938. Betty's father Usang Ipoi passed away 8th May 2012 and Betty's mother Cecelia Liwan passed away 8th May 2020. Born on the same day and died the same day of the month. As it happens 8th May is also VE Day.

The Brigadier General of Malaysian Air Force came to see me one day and asked if he could bring his men and climb the Pinnacles. He did not need to spend any nights at Camp 5 where we had accommodation for tourists as he had a Sikorsky Helicopter. I agreed to his plan providing he did not use the park for military exercises. We had a couple of large concrete helicopter landing pads so he popped in with his men. He was a charming man who loved to see his men having a good time. Besides borrowing his Sikorsky and flying around the park we had some great nights with much beer and singing at the park guest house. One night we ran out of beer, no problem he said I will send the Sikorsky back to his base at Labuan Island for some crates-which he did. Being such a nice guy I arranged to borrow his Sikorsky and his pilots to transport materials to Camp 3 which was a camp going up to the summit. A great team of maintenance staff climbed to Camp 3 and built a good camp with water supply and toilet. I inspected it later with a plan and measurements during a trip to the summit with Matt Kirby. Matt is one of those exceptional cave explorers who always seem to have a smile on his face even under difficult circumstances. We explored together during the Mulu 1988 expedition and from then onwards Mulu became an obsession with him leading the majority of expeditions to this remarkable karst area. He also compiled the majority of the Mulu Caves Project reports. Mulu seemed to become his caving life's work.

When the road was under construction I noticed that they were building a large concrete structure opposite the park headquarters. When I asked what they were doing the answer was a road bridge into the park. I put a stop to this and we built a suspension wire bridge in keeping with the surroundings. No more did we need a boat to cross the river.

The airport was opened and the visitor numbers increased to 12,000 in 1992 but the adventure of the river transport had ceased. Visitors came for two days and one night instead of staying for many days. The problem with the airport was the evening flight back to Miri. It worried me that two million free tailed bats flew from Deer Cave over the area in the evening with the chance of a collision with an aircraft. I wrote via our head office to inform Malaysian Airlines of the potential danger and they stopped the evening flights. Possibly Gunung Mulu National Park is the only one in the world to halt flights in order to protect its bats.

Regular VIP visits proved to be time consuming as I was assigned to be the guide. Ministers, Ambassadors, Princess of Thailand and Prince of Saudi Arabia to name but a few all came to visit. Caving friends from the

UK were constantly arriving and staying at our quarters which proved to be entertaining. Over the years we also found the time to explore and map a few caves.

Inspecting the trails was always hard work in Mulu. The rain when it started was torrential bringing out the leeches in profusion. Getting to Camp 5 was a one hour journey upstream on the Melinau River, where in dry weather one needed to push the boat through the rapids. An 8 kilometre trail led to the camp which was one hut which could sleep 15 people. This was situated in the Melinau Gorge, a beautiful spot overlooking the dramatic cliffs of Benarat. Upstream was a deep and wide swimming pool where I spent my free time at Camp 5 and also explored the area for caves. The trail to view the pinnacles involves a three hour long 800 metre climb where I spent time with park guides fixing ropes and aluminium ladders. In the tropics the karst becomes unstable quickly due to the high rainfall of 4000 to 5000 mm per year so rock anchors and trees often collapse. We fixed ring bolts directly into the rock to hang ladders and ropes safely as I had brought a few with me from England and a bolting kit.

The 11.3 kilometre Head Hunters trail started from the other side of the river so we built a rope bridge as, when in flood, no one could cross. This is a wonderful trail which terminates at the Terikan River where it joins the Medalam River. Downstream by boat reached the Mentawai River where we had a Ranger Station. In later years we completed some renovation and tourist accommodation which unfortunately is never now used. Tourists tend to travel down the Medalam and spend a night at a Longhouse before a road trip to Limbang Town. A few years later we constructed a small day shelter near the Terikan River and upgraded the trail.

The trail to the summit was much harder, 24 kilometre uphill climb to 2,377 metres above sea level. I had three camps along this trail, Camp 1, 3 and 4, all basic timber construction with water, toilets and cooking area. Camp 4 was at a high altitude and cold at night but with fascinating stunted vegetation.

Braken, our head warden from Miri, wanted to build a pathway from headquarters to Clearwater Cave passing Wind Cave. The object was to reduce the river traffic as the Clearwater jetty was becoming congested. We set off surveying a route but came to a sudden stop at the high steep cliff of Batu Bungan. Marking a trail up the steep jungle clad hill we found a cave which amazingly passed directly through the cliff-which was adorned with nice formations and Moon Milk. Moon Milk is a calcite formation which has undergone biological transformation into a soft milky substance which covers the walls of the cave. It finally took two years to obtain a tender for construction of the trail which proved to be very popular with visitors but sadly has now partially been closed due to lack of maintenance. The legend of the Batu Bungan cliff revolves around a young lady called Bungan who lived in a cave in the cliff. A hunter made many attempts to reach the lady but failed so, in frustration, he killed her with a poison dart; thus the cliff is said to be haunted.

The main problem in Mulu arose when they began to construct the airport and the luxury hotel. The hotel was constructed on the old park base camp just outside the park boundary. Unfortunately this area flooded during periods of heavy rain owing to the backing up of the Tutoh River. Our new headquarters was actually inside the park further upstream-supposedly above flood level. This proved not to be true as the water was knee deep at times. A road was constructed which led from the airport to the hotel with a spur road leading to the park. The Berawan claimed that this impeded on their land and no compensation was given as officially it was claimed as state land. This increased the animosity felt by the local population and would eventually lead to trouble.

Unfortunately, long term planning for Gunung Mulu National Park and the surrounding area had not been fully appreciated with a Penan Longhouse called Batu Bungan constructed up river opposite the park boundary. The Penan were originally nomadic but had settled near the boundary. This was a great advantage for the Penan as the government built them a Longhouse which they accidently burnt down. They built them another one which they accidently burnt down once more. The third one is still standing but too small so the area is a conglomerate of wooden huts with much pollution and an alarming growing number of settlers, many from outside the region. The government built them an agricultural centre, a clinic and a school and they have work at the park and from tourists if they need to earn money. The other advantage is the protected primary rainforest is opposite the village so open for hunting and gathering.

To establish a totally protected area the local community have to officially be informed of the intention. This means in many instances that privileges to hunt and gather are granted and the intended boundary would need to be altered. The privilege is stipulated for certain areas and only pig and deer can be killed for home consumption. The totally protected area does not become total, only partial, causing huge problems for enforcement. As there was no enforcement at the park the Penan could kill anything they wanted, anywhere they wanted. This has caused the wildlife to disappear. Tourists ask why there is an absence of wildlife expecting to see animals in profusion. The answer is not easy to give.

Presently the Melinau River is highly polluted as there is no waste management-so rubbish and sewage finishes up in the river, not ideal for the local community and this problem will only escalate. Berawan families built houses and farms all along the river bank and also along the road so the population at this point in time is growing to over 800 people without electricity-except their own small generators and without clean water or efficient sewage. Hopefully this will change in future years.

Even the Royal Geographical Society expedition in 1974 to 1975 encouraged the Penan with little thought to future trends. A handful of people poaching would not be a problem but in the hundreds it is.

The school was a big help as the children could now be educated but on reaching the age of 12 years they had to move to the boarding senior school at Long Panai situated a longboat ride away downstream, almost a four hour journey. They did not want to leave their parents so no one went to the senior school. I received a donation to help the school and bought school clothes, bags and books for the children. Also I managed to hire a boat to take 16 children down to the Long Panai senior school. Many did not like being away from their home and deserted but, over the years, it worked and now the children have a higher education. This was a step in the right direction for the Penan which I hoped would give the next generation a chance to work for a living and halt the poaching.

To sum up; the park boundary should have included the land on the true right bank of the Melinau River as far as the international boundary of Brunei Darussalam-and also a section on the left bank. The park headquarters could have been built at the junction of the Melinau and Tutoh Rivers. The airport, the houses and farms, also the Penan village are now all in the wrong place-directly threatening the conservation integrity of the park. By 1992 and today there is little that can be done to mitigate the problems.

Working on conservation over the years it became obvious that, although much had been written on conservation of cave and karst protected areas, the real answer was enforcement of the laws governing the area. Park Rangers do not want to arrest locals as they live there as a part of the community. The only way to achieve the aim of conservation is the total involvement and commitment of the local communities-so that the benefits arising from tourism come to them. In this way awareness programs might work to convince the local community to safeguard the park as it is within their own interest. Sadly the above has not yet been achieved at Mulu.

Poaching of bird nests became a huge problem growing over the years. The nests of *Aerodramus maximus*, the Black nest Swiftlet is valuable for bird nest soup. We captured a few poachers but it proved impossible to find the groups in the forest and the caves where it was easy to hide. The introduction of field force police did not solve the problem but I managed to keep it under control to just a few groups with a spy network of informers. This did make me unpopular with some of the community which occasionally resulted in physical threats.

Serious trouble eventually did break out as a family built a house in the grounds leased to the hotel-claiming it as their land under Native Customary Rights and no one seemed to stop them. As this was outside the park boundary it was none of our business. Unfortunately the owners of the house blockaded the jetty where the hotel landed their fuel supplies. The police were called and a few of the local community were arrested. This led to a general strike of all the people working at the park and the tour operator's guides. They also blockaded the road leading from the airport to the hotel. The area became crowded with police field force officers heavily armed. Negotiations came to nothing as the arrested were incarcerated for six days. Eventually an agreement was reached, the arrested released, the building purchased from the family and the staff came back to work.

Trouble did not end there as, a while later, the Bat Observatory at Deer Cave was burnt to the ground along with my generator house at Wind Cave. The generator was completely destroyed. A guide put the fire out in the Clearwater Generator house and saved the generator from serious damage but the repercussions proved to be serious. We could not understand that the park was attacked as many of the local community owed their existence to working at the park and the tourism business. Once more the park was inundated with police and field force all trying to find the guilty parties. Arrests were made but weeks passed before there was a breakthrough with a group being charged. At the hearing the judge dismissed the case and we were once more almost back to the normal peaceful state of affairs.

Along with a maintenance team I managed to run a heavy duty 3 phase cable from the Clearwater Generator to Wind Cave to get the lights working once more, a hard job.

A great day arrived on the 2nd of April 1994 when Betty gave birth to my daughter Racheal Saliah, the first girl in my family for a generation. She was on a light aeroplane heading to Mulu within a few days of her birth.

Headquarters Mulu, Betty, Racheal and the author. Photo: Matt Kirby

I started work, under instruction from head office, to assist the Raid Gauloises organisation team as they wanted to include Mulu and the caves as a final section of the five day race. The race involved a trek up

Sarawak's highest mountain, white-water rafting and mountain bike riding. As the teams would be experienced in adventure pursuits which included caves I designed a trip from the Tutoh River across the forest to Deer Cave. From there they would traverse the Garden of Eden and into the splendours of Green Cave to emerge at the Melinau Paku Valley. Crossing the valley, a pathway through the jungle would lead to Barangs entrance into Clearwater Cave. A complex series of river and dry passages leads to the show cave entrance where they would then paddle canoes to the finish line at park headquarters. It was a long but very exciting underground journey and all the teams were fully kitted out with all the necessary caving and safety equipment. Everyone agreed on the plan and it was arranged that, on my return from the UK, I would brief all the teams on the route.

The trouble was that my application for another contract was surprisingly dismissed. We left Mulu with a heavy heart but the Minister phoned me in England with the information that they had appealed against the decision.

Betty was not very impressed with England-probably due to the cold! Driving around Derbyshire she asked what all the little white things were in the fields. When I said sheep she asked if they were dangerous. Racheal was a big hit with my dear parents being the only girl and it was wonderful to see mother and father again after almost three years.

We returned back to Sarawak and I phoned the Minister and Director of Forests and they offered me another contract but only for one year. The briefing of the Raid Gauloises teams went well and the State Secretary told me they had received a bad police report. I thought that this was most probably due to the problems at Mulu as the police had asked me who the ring leader was; being a foreigner they thought it was me. Why, after I had spent so much time and effort in upgrading and installing the generators, I have no idea. The Raid Gauloises was a huge success and the Gunung Mulu park guides won. I inspected the caves later and there was no rubbish found. The going away party at a hotel in Miri was a little wild with most people being thrown into the swimming pool including me and the hotel manager.

From then onwards I managed to obtain a one year contract. This was fine as it gave me the opportunity to travel back to the UK every year free of charge with Betty and my children. One such year my son Jeff came with us and, stepping outside at Manchester Airport he looked up and marvelled at the way the English produced air conditioning from the sky.

As the years passed quickly I had much fun over many months completely re wiring Clearwater Cave with the help of a relative of Betty who happened to be an electrician. I managed to design and install a galvanised pipe line from the river for diesel to be pumped directly from the fuel boat to huge steel storage tanks. I had the opportunity to design and build the control panels which saved a great deal of time and completed designing, building and installing new control panels in Langs Cave.

As visitor numbers increased, Camp 5 became inadequate in size to cater for the sleeping of greater numbers. We decided to build a much larger building with kitchen and toilets. The contract was tendered out and the winner of the contract came to see me asking about boats to Camp 5. When I explained that boats could only go as far as Kuala Berar, approximately a one and a half hour upstream boat journey then a walk of 8 kilometres, he fell silent and went pale. He asked me 'why was there a jetty on the drawings'. It seemed strange to me that he had tendered for the contract without a site visit. Transporting all the materials to Camp 5 was incredibly difficult and helicopters were far too expensive. Eventually he constructed an 8 kilometre long wooden rail system. It took over a year to complete the facility and not long afterwards the poor man passed away.

Working at Clearwater one day I was wearing my bright orange caving suit. Two young English cavers were around enjoying adventure caving and had just returned from Camp 5 where they had been climbing up to the Pinnacles. One sat on the jetty looking sick, he said he was OK but jumped in the river, swam to the other side and disappeared into the forest. I asked his friend who informed me that he had a very high temperature. I had a radio transmitter and called out a team of guides and maintenance staff to search for the missing lad. Travelling upstream in a longboat we spotted him sat in a tourist boat. They had seen him sitting by the river and picked him up. We took him to the tourist quarters where our medical orderly pronounced him to have a dangerously high temperature. He was hallucinating with the fever so we sent him to hospital in Marudi where he was diagnosed as having Leptospirosis, known in England as Weil's disease. The disease is derived from contact with urine, blood, tissue of animals or rodents that are infected with the bacteria. In England it is usually from rats. This disease had occurred previously with cave explorers in Mulu so must derive from contaminated water or wounds in the caves probably from bats. We had a general guideline for cavers never to drink the cave water and clean wounds with antiseptic. The lad recovered with antibiotic treatment and many years later I was attending a dinner in the Yorkshire Dales where I had given a lecture on our expedition to New Britain Papua New Guinea which took place in 2006. A caver came over to me and said I had saved his life as he was the young man who almost died in Mulu all those years ago.

Rudang Gallery, Blackrock Cave. Photo: Jerry Wooldridge

Looking back over the many years I spent working in the Gunung Mulu

National Park the problems could be resolved with thought and planning. Identify the problem and work out a solution, find the funding and materials to accomplish the task. This proved to be great fun and highly satisfying.

We tend to have a fear of dense tropical jungles but living in one for many years is not as bad as it sounds. The vast species of trees and plants are without doubt fascinating but there are dangers to contend with.

The crocodiles found in the Borneo Rivers account for a few lives as they are clever creatures and wait, watching people bathing or swimming, then, given the opportunity-strike. They deposit the prey beneath river undergrowth to eat later. My young son Jeff had a narrow escape when swimming with his friends in the Tutoh River. Luckily my father-in-law Usang Ipoi spotted the crocodile and shouted to the boys to vacate the river fast. He took a shot with his shot gun and scared the crocodile away. Crocodiles are now a protected species so the numbers are on the rise.

Snakes are another problem which claim lives. The viper coils itself around a tree branch and is well camouflaged so it is easy to use one as a hand hold. Descending from Green Cave with Matt Kirby one day using small trees as hand holds, I missed a very large viper by inches, it looked at me straight in the eyes and told me to go away.

Stinging plants are also a problem as are rattan plants. Rattan has fish hook needles pointing out to scratch and tear the skin. Poisonous insects are always hanging around to bite you along with leeches. One of my friends wore ladies' nylons as the leeches could not get through the tight nylon weave. A major problem with leeches, they tend to crawl up your legs and attached themselves to the private parts. I collapsed one late night with a swollen and painful arm. The doctor pronounced a bite from a poisonous centipede.

Animals are not a problem as they tend to keep away otherwise they are likely to get shot. The answer is, keep vigilant.

2

Karst Management Unit, Sarawak

'Three things in human life are important; the first is to be kind; the second is to be kind; and the third is to be kind.'

Henry James

Development and upgrading at Gunung Mulu National Park had made considerable progress over the past few years and the Forest Department often asked me to look at other karst areas in the state. It was suggested that we form a Karst Management Unit of interested individuals from National Parks and Wildlife Department. To clarify; karst is just the geological technical term for limestone areas where the majority of the rainwater sinks underground. These areas in the tropics are of great importance for their high biological diversity but sadly limestone is a desired product; so many areas are destroyed by quarrying activities, along with the biology and caves contained within those areas.

There were hundreds of caves, few of them mapped and it was possible to work on conservation of these other karst areas and form national parks or nature reserves. It was estimated that we had 14 separate karst areas within the state. I produced the guidelines and the documentation for research-including cave exploration expeditions-which was accepted. It was important that all research information including cave exploration reports and cave surveys and the raw data were supplied to the Forest Department-as much information had been lost in the past. The old survey data from previous cave exploration expeditions was lost for all time. I spent years trying to collect it all together and inputting with data into a survey programme called 'Compass'. I stored all the data in box files which were deposited in the office library. Years later I tried to retrieve them but strangely they had disappeared.

I was now the proud possessor of a basic laptop with email which made life much more convenient. It also holds my veritable mountain of files.

The bible of Sarawak and Sabah caves is still, up to this day, the mammoth work of Dr G.E. Wilford in his book 'The Geology of Sarawak and Sabah Caves' dated 1964. If anyone needed to conduct ground work on the caves, this was the essential book to read. I spent many fascinating hours reading and taking notes. His map below gave me an indication of the karst areas of Sarawak and caves mapped prior to 1964 so the book allowed me to formulate ideas on where to start.

There were still numerous projects at Mulu that required my attention. The Mulu guides still had the same problems with being on a free lance basis. It was stipulated in the park guidelines that all visitors should be accompanied by a park guide. I fully supported this as it gave employment for the local community so reducing their cultural desire to go killing wild animals.

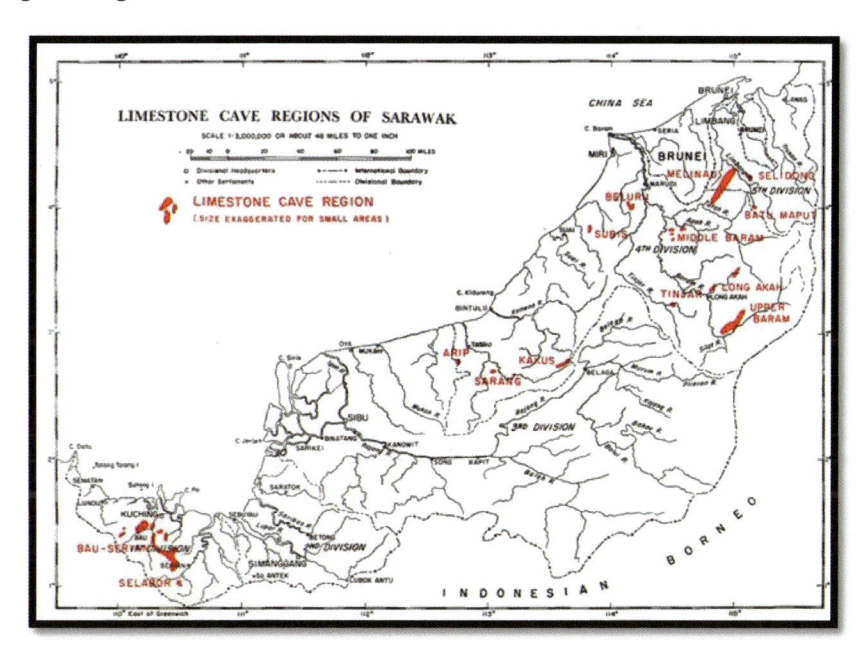

Map of the karst regions from Dr G.E.Wilford

With the assistance of Garry Tay we managed to offer them permanent employment as official park guides. I organised a roster which gave every guide a weekly change of scenery. One week stationed at Deer and Lang's Cave, which gave them the morning free, one week at Wind Cave and Clearwater, which gave them the afternoon free, one week at Camp 5 which enabled them to claim an extra allowance and a few on standby for Summit trips and Adventure Caving. They seemed happy with the arrangements and I managed to recruit a few more keen individuals giving them training on caving techniques including single rope techniques. I had in the past kitted out guides with caving equipment but most of it went missing and also someone broke into the caving store and helped themselves to karabiners. Probably the men who went bird nesting as caving equipment is useful for climbing the walls of the caves in order to steal the nests.

Another problem was the old antiquated system of all tourists requiring a permit from the Resident Office in Miri and a Police permit to visit Mulu. This was originally devised to prevent the expansion of the coastal Iban groups for moving up river into the Orang Ulu group areas. It was now totally unnecessary and prevented an increase in tourist numbers. Garry Tay in his efficient self managed to sort the problem out so no permits were necessary.

As visitor numbers increased yearly we were still short of guides and a number of them and other competent individuals requested permission to escort visitors in the park without being accompanied by an official Park guide. I therefore set up a training scheme with training manuals which I held on a yearly basis awarding Free Lance Guides with a certificate and licence to guide visitors in the national park. This worked well and was later extended into specialised modules with the support of the Ministry of Tourism for all national parks in the state. It was felt important that senior national park officers knew something about caves and caving so I started training courses on speleology, single rope techniques, cave rescue and mapping. Wind Cave and Fairy Cave in the Bau karst region near Kuching was an ideal place to start.

The work in Mulu was continuous, with Officers in Charge normally lasting no more than two years. Many came straight through from the Forest Department so had little experience of National Parks and the problems associated with tourism. The ordinance governing totally protected areas included concession management. This in effect meant privatisation of the tourism products under a management agreement with the Forest Department who would still be responsible for conservation and enforcement. We had visits from the Australians from

Jenolan Caves who of course wanted to be involved in the management of the park. They had not been invited by the Forest Department but were privately funded. Their report did not tell us anything we did not know but as the tourist figures were increasing year by year the basic idea was a good one. We even had a visit from the leader of the Royal Geographical Society Expedition from way back in 1974 and 1975. They also submitted a proposal for management which was again privately funded. It was something to be seriously considered in the future.

VIP visits to Mulu seemed to be a major attraction as the authorities were rightly proud of the caves and karst area; we even won an award as the best tourist attraction. There were many funny stories as, during the troubles, I was guiding an ambassador who did not turn up from the hotel for his trip to Wind Cave and Clearwater. I found out via devious means that the police had arrested his boat driver. We obtained a boat and everywhere we went we were followed by a special police officer. The ambassador wanted to swim in the crystal clear deep water at the Clearwater resurgence. We stripped off and dived in again to be followed closely by the policeman in his knickers.

The Princess of Thailand wanted to trek through the forest in the pouring rain. Her huge retinue, the men dressed in suits and the ladies in smart dresses plus numerous armed police had to follow. When we returned they all parted each side of us in the jungle, they were covered in mud and bowed low as we passed. Now I know what it feels like to be a Royal.

A helicopter landed at park headquarters one day and out stepped two Princes of Saudi Arabia accompanied by the Sarawak State Secretary. They wanted to see Deer Cave but not by walking along the plank walk to get there. We had a helicopter landing pad close by Deer Cave so they decided to fly. The Officer in Charge of the park was nowhere to be seen along with the four Forest Guards; luckily I had Garry Tay with me so we both raced up to Deer Cave to show them around.

During the Raid Gauloises trips one of the senior ministers from West Malaysia turned up to see Sarawak Chamber before he retired. It was my job to guide him and there was no way he would take no for an answer. His retinue included government officials and some ladies. This was going to be a difficult job as none had ever been in a cave before. I managed to use all the parks lamps and helmets and set off for the three hour walk through the jungle to the entrance with a few of my park guides. Some of the retinue turned back when they saw the underground river which caused problems as a guide had to accompany them back out of the cave and through the jungle. The old gentleman VIP made it all the way to the chamber and sat on a boulder with all our lights extinguished while he prayed in the dark. The journey back was a nightmare as his lightweight shoes fell apart so I tied them together with string. It was around midnight when we finally arrived back at park headquarters.

I was guiding his Excellency the High Commissioner for Argentina one day and just could not resist asking him how he got on with the British; his answer was a good one.

Cave of the Winds (usually referred to as Wind Cave) continued to be a huge problem with the plank walk which required daily repairs. I inspected all the timber and came to the conclusion that it required scrapping and rebuilding as the main supports had degraded. My report asked for 15 ton of timbers as a temporary measure. When it did not appear I threatened to close the cave on safety grounds. The timbers appeared and we could carry out repairs. I accused the contractor of cheating as the timbers were not as stipulated in the original building contract of iron wood known as Belian, but Meranti wood had been used. I was then informed that special permission had been granted to the contractor as the King was arriving to see the cave so they only had three months to complete the installation and Belian was not available. My recommendation and design was a walkway built of sustainable materials, galvanised steel and concrete, the same for Clearwater which included a bridge across the river and modern lighting. My design was rejected in favour of using timber. It seemed that I was employed to give advice which was often ignored.

The building consultants submitted the plans but the Chief Minister wanted to visit Cave of the Winds. I recommended that we looked at Clearwater as it was safer. This was ignored and the retinue numbered a hundred. I gave instructions for groups of not more than ten persons at any one time but over 50 people all piled onto the briefing platform. Just as I was giving a briefing about the cave the platform collapsed, sending everyone sprawling. The Chief Minister held his ground as I explained that the walkway was going to be rebuilt of timber. The Chief Minister took one look at it and asked

'Why can't you build it of steel and concrete?'

'Just what I wanted to do but it was dismissed' I said.

'We will see about that' said the Chief.

In the Kings' Chamber he grabbed the Director of Forests and told him he had discussed it with me and thought it should be built with sustainable materials.

'Too expensive' said the Director.

'There is no compromise for safety, bypass the system and see me if you have a problem' said the Chief.

So in the end I got my own way. I was accused of sawing off the leg supports of the platform. I pleaded not guilty.

Of course problems at the park persisted with a number of tragic accidents. A falling tree at Clearwater cave entrance area killed a few tourists. I was asked to chop down dangerous trees but in the rainforest all trees are potentially dangerous. During a storm they regularly crash to the ground, it's a natural part of rainforest dynamics. Another nasty tragic accident occurred during a descent down to Camp 5 from the Pinnacles. A man slipped and fell spearing his side via a short tree branch. It was dark and by the time help arrived it was too late to save his life.

A complaint from a tourist claimed the Pinnacle Trail was dangerous which of course it is. I had spent much time fixing ropes to rock anchors which I had to purchase out of my own pocket as the park did not have any-so we all knew that tourists needed to take great care. I was instructed to inspect the trail so spent a nice day on my own climbing the trail and taking photographs. Of course the trail was just the same, no damage or worn ropes. It appeared that some of the tour operators were not briefing their clients on the severity of the Pinnacle Trail as many thought it was just a short walk-when in fact it was an 800 metre climb using fixed ladders and ropes.

A report originating from our guides indicated that the Terikan River that emerges from the Terikan Cave was muddy. This is just off the Head Hunters' trail which leads from Camp 5 to the Medalam River. Tourists can then travel downstream by boat and then road transport to Limbang town. This river is normally crystal clear so it was suspected that illegal logging had commenced within the park and the streams feeding the caves were polluted. The cave is situated at the north side of Gunung Benarat. I recruited one of our highly experienced guides called Lawai Kumpang and a few Forest Guards. The wonderful thing about our guides is their excellence in the forest and all of them brilliant in caves.

Via Limbang town we travelled on logging roads to our basic lodgings at the Goodwood logging camp where we spent a few days investigating the possibilities of illegal logging. The river was clear and no logging encroachments could be found. The pollution of the underground river could have occurred within the cave system via a collapse. The only thing found was illegal hunting trails established by poachers. Being in the area did give me a chance of discussing the possibility of establishing Gunung Buda as a national park with Pengulu Siga from Kuala Medalam Longhouse. The title Pengulu means he is a chief of his area. His main concern was the father of one of my park guides, he had established himself in a small house near Turtle Cave where he was bird nesting.

Sometime later I was asked to present a paper at the Malaysian Adventure Festival in Kuala Lumpur, the capital. The organiser promised to pay for our air fare but of course I never received his payment. Again I took with me Lawai and after the festival we were asked by the manager of the Tekam Plantation Resort to look at the cave on their plantation land called Gua Gelanggi. They wanted to develop it as a show cave and, as he promised to provide transport and accommodation, we agreed. We mapped the cave and I provided them at a later date with a development proposal with structural drawings and electrical layout.

While in the area it gave me a chance to visit a few caves in the state of Perlis, the Setul Karst Area with Hymeir Kamarudin. We stayed with him and his wife and went exploring. We took my son Jeff with us down a nice stream cave but half way through the stream started to rise dramatically. There was only one alternative, get out as fast as we could. The stream started to rise in ferocity and as Jeff was only small the waves were coming over his head. We made it to daylight but only just. Hymeir along with Liz Price were both leading lights in Malaysian caving and members of the Malaysian Nature Society. It was hoped that one day they and their caving friends could organise an expedition to Mulu but sadly it never occurred.

Filming at the park was a continuous problem as so many international film units arrived over an average yearly period. They were all supposed to supply us with copies of the films but they never arrived. The Koreans wanted to film the show caves along the walkways but I had to insist they did not set up their electrical generator inside the caves. Of course they ignored the request and I even found a climber climbing up the walls. A notable one was Rob Franklin who arrived with his film crew for a Malaysian TV special. The concept was a youngster who wished to experience a high adrenaline adventure; this one was caving in the longest cave in Asia which happened to be Clearwater. We had fun on the cliff at the rear of the hotel teaching him single rope techniques and some filming in Racer Cave. The big trip was going to be a long and hard one which involved trekking for two hours through the jungle along the Melinau Paku Valley to a small entrance near the river. This led to the magnificent Clearwater 3 river passage where we spent considerable time filming. Just prior to the downstream sump pool, a climb up to the huge high level passages and a final descent to Battleship Passage-so named due to the stalagmites on a huge rock pointing diagonally towards the enemy. Joining the main Clearwater River to the entrance took many more filming hours. We emerged around 3 am the following day. The film was very well received, a great credit to Rob and his team.

I was then surprisingly asked to transfer to our head office in Miri leaving my beloved Mulu behind. This seemed strange as I had been initially employed specifically for Gunung Mulu development. Braken had been transferred to Kuching and Simon Sandi was now my boss. I had explored caves with Simon and he was very interested in the idea of the Karst Management Unit.

This meant moving everything I had, plus the family, into a government house. It also meant finding a car. The house the resident office allocated was an old 1946 built wooded single storey house up on the hill in Tanjong, Miri. A beautiful setting surrounded by jungle but the house was falling to pieces, Betty spent ages trying to clean the place up and my friends from the Public Works Department renovated the place for us. I then purchased a cheap car and managed to obtain some furniture. This was a wonderful place as, most weekends, we had crazy parties with Betty's relatives and the neighbours never complained. Jeff and Racheal could also attend school in Miri. The house had some historical significance as one day I saw a European in my garden taking photographs of my house. It appeared that he lived there during the

British administration. He played me a sound recording of a woman singing to him many years ago at our Longhouse in Long Terawan and wanted to meet this beautiful woman again. Betty's Cousin Martha recognised the voice of her dead mother and had to tell the gentleman that she was no longer alive.

I moved into my office which happened to be the store room with no windows, rather dismal but I could now start work on collating a library of cave, geology, karst books and papers; also ordering caving equipment and planning expeditions to other karst areas. One major task was to continue with collation of all the known information on Gunung Mulu National Park. This included scientific reports and cave expedition reports. A great problem encountered was that the cave mapping data with notes, initial drawings, survey station numbers and measurements had to be found. This information was not in one place but spread around different cave explorers. After many months the collection grew and all I had to do now was to collate it into some semblance of order, including entering all the measurements into a cave mapping software programme called Compass.

The redevepment of Wind and Clearwater Caves proved difficult as of course the caves are a three dimensional maze. At a meeting with the State Secretary looking at the drawings he suggested more pathways of concrete. I redesigned the drawings and submitted them-which was accepted by the consultants. Each steel support needed cutting to the actual size, which the contractor did not do, thus some supports were too high. The result was the cutting of each support and then re-welding to the correct height, a mammoth task. I built an experimental control panel using solid state relay and a cut-off timer which worked well in order for the electricians to rewire the cave lighting. It was essential that the lights along any given section of cave passage must switch off after a short time as if they are left on for a long period, lampa flora grew on the formations due to the photon emissions. This was green algae which can only be removed with mild bleach. At a meeting in Kuching with the State Secretary and State Planning Unit we received much criticism as the cave looked like a building site which was exactly what it was.

3
Caves and karst regions of Sarawak

'Even if you are on the right track, you'll get run over if you just sit there.'

Will Rogers

I had many expeditions to different karst areas surveying the caves wherever possible over a number of years with major help from Mulu National Park guides, all good cavers with experience.

One particular area of interest was the Kakus region which we knew little about. The area was briefly mentioned in Wilford's book with a map. Sixteen caves had been marked by his assistant but none mapped and no descriptions. The karst was a thin strip approximately sixteen kilometres in length with access along logging roads via a 4 wheel drive vehicle. This limestone outcrop is composed of Miocene limestone, the same as Gunung Mulu as it lies in a direct line north east to south west. It is massively bedded with some sandstone, marl and shale and the Kakus River flows along the southern flanks. From Bintalu town situated on the coast it was around 83 kilometres south east but twice as long along the rough tracks. The caves are the habitats of the Black nested swiftlet, *Aerodromes maximus*. The nests of course are highly prized for bird nest soup sold to Chinese traders. The problems with caves where nests are collected are the squabbles appertaining to who has the right to collect these nests. A licence needs to be issued by the National Parks and Wildlife Department and in this case two families claimed a cave with two entrances. One family claimed that the entrance higher up the mountain was theirs as it was a separate cave and not connected to the lower entrance. The opposing family claimed the cave was all one and therefore they had the right to the licence. It was up to us to map both caves to decide who was right.

In October of 1995 I had two small teams in order to map this cave which was called Lubang Bedawa situated in the western sector of the limestone. These old inactive caves are short in length and had been formed originally by rivers flowing from the high ground northwest to southeast. This one was large in size but full of guano or better known as swiftlet shit. I had managed to recruit a couple of guides from Mulu, Peter Tepun and Philip Lawing both very competent in caves so they headed off to the upper entrance with Braken and my team to the lower entrance with Simon.

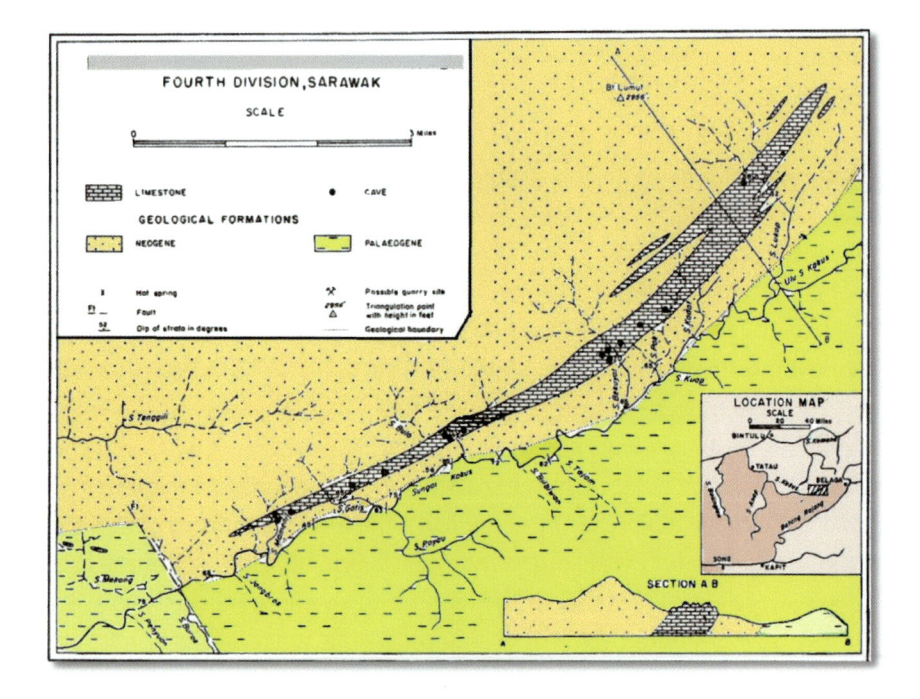

The Kakus limestone area

The lower entrance descended steeply to a side passage and a chamber 10 metres high. As expected the other team's lights appeared up in the roof proving the cave was one and not two. We completed the mapping and added further passage on another day but the cave length was only 500 metres, short by Mulu standards. The map settled the dispute as regards the licence to collect nests.

Another dispute occurred a few years later when two families claimed the same cave called Lubang Danau. I mapped this cave with Victor the ex Officer in Charge of Mulu National Park when I first started the job. This one was even more appalling than the first cave with two pitches of 6 and 12 metres where we used ropes to descend into a 15 metre high passage and a lake. The nesting areas were high above and the only access was to swim the lake to gain passages leading upwards beyond.

The problem was that the lake was composed more of swiftlet shit than water and swimming across was a nightmare as we emerged covered from head to toe in very smelly guano. The passage beyond was equally disgusting with a low crawl passage heading upwards full of large cave insects feeding off the guano with the insects tending to slide down our necks. We reached a large ledge with a 10 metre long bamboo pole leading across to a ledge on the far side. This was 20 metres above the smelly lake and one of the family members who claimed the cave was his proceeded to balance across to the far side. We of course declined his kind invitation to follow. To make it more confusing the two families had different names for sections of the cave.

My map was presented to a judge at a court case between the two families but although called to give evidence I was out of the country at the time.

It became obvious that to constitute a Nature Reserve at Kakus would be fraught with difficulties due to all the various families claiming bird nesting rights to the caves. A Nature Reserve means that the land would become State Land which would not be acceptable to the local community so the plan was dropped.

The coastal Similajau National Park lies to the east of Bintalu and has magnificent coastal scenery and forest. A report came through that caves had been registered for bird nest collecting in a palm oil estate on the fringes of the park. Using a four wheel drive truck along rough tracks we eventually found a hut which was a guard post. The hut was close to five cave entrances with no limestone apparent on the surface. The first was called Bat Cave for obvious reasons. We entered via an 8 metre pitch where we abseiled down into a large passage leading to a walled off continuation. Another three entrances were found and the cave mapped for 274 metres, again only short but it appeared to be formed at the top of the limestone sequence having a chalk and shale roof. Another cave close by called Gua Semba was similar but only 54 metres long. Besides bats the caves were the habitat of the White nested swiftlet *Aerodromes fuciphagus*. These nests are exceptionally valuable as they are produced purely by spittle from the bird whereas the Black nest swiftlet also use feathers as a part of the nest. The black nest can fetch as much as £400 per kilo untreated in its raw state while the white nest £1,000 or more depending on the market price. As the area was a palm oil estate there was nothing I could do with regard to conservation of the caves.

The Middle Baram karst had again been noted by Wilford and bird nesting licenses had been issued to various families. Again the caves were the habitats of the White-nest Swiflet *A. fuciphagus* so the nests were extremely valuable. This area is one of the most interesting and is situated close to a town on the Baram River called Long Lama. Yet the limestone seemed to be in over 28 separate outcrops surrounded by jungle spread over a distance of 126 square kilometres. It had been completely overlooked by cave explorers as Mulu had taken pride of place. This looked like an ideal area for a new national park or large nature reserve. The problem at this time was access. The road from Miri to the junction leading to Long Lama was good but from then on it was a rocky and very rough track that regularly flooded in places. The distance was around 135 kilometres.

The totally protected areas of northern Sarawak

There was no bridge across the Baram River just a ferry boat. This road has at last been up graded and a bridge built across the river. We made arrangements for a meeting with the licence holders at Long Laput on the opposite side of the river facing the limestone areas. The group of people at Long Laput are Kayan's, an Orang Ulu group as distinct from the Iban people who originally were settled around the coastal regions of Sarawak. Orang Ulu means Up River People, the same as the Berawan from Mulu. The meeting was well attended and I introduced myself as being married to Betty Usang her father being Usang Ipoi, a well known man in the area. I eventually became like one of the family. We explained why we thought the area needed total protection under the National Parks and Nature Reserve Ordinance as the Gading quarry to the north was expanding with the ultimate loss of the caves and the swiftlets.

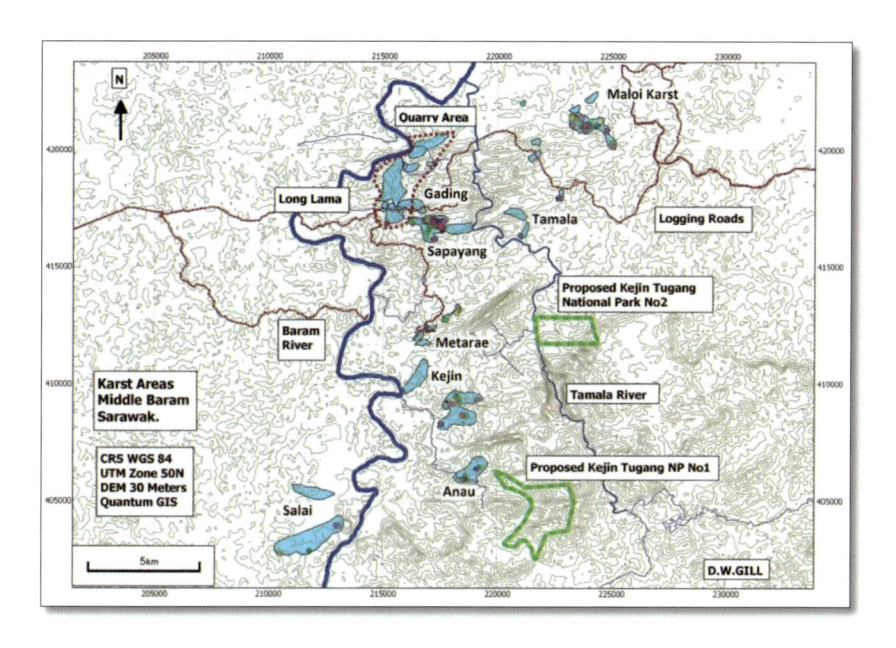

The limestone areas of the Middle Baram, Sarawak

The danger of oil palm estates destroying the natural forest cover where the birds feed was also a substantial threat. It was estimated they had over 100 cave entrances in the region but bygone poaching had reduced the birds to 15,000 breeding pairs. Major efforts had been made to prevent poachers by building guard houses with an armed guard. A few of the highly productive caves had been fitted with massive steel locked gates with access gratings for the birds.

Agricultural privileges are awarded to the local communities to farm their secondary forest, cultivate rubber and fruit gardens, these are designated as Native Community Registered land areas.

I was lucky to have with me the world's leading expert on these amazing echo locating birds, Lim Chan Koon was undertaking his PhD and knew more about their breeding habits than anyone else. The meeting agreed with the basic proposal but with some misgivings and permission was given for me to map some of the major caves.

These cave swiftlets have the ability to navigate up to two kilometres from a cave entrance through the cave in complete darkness by echo locating. While camping underground the click clicking noises of the thousands of birds flying past can become quite hypnotic. As we are not supposed to be there, they can collide and land in the soup.

It was fascinating to listen to Lim on his studies concerning the sustainability of taking swiftlet nests. The first nest could be taken before the swiftlet laid its egg, usually only one egg. The bird would then make another nest but it was essential that this was NOT taken as the bird would be unable to build another one. When the hatchling flew away, this nest could then be taken. On the second breeding season the nest would be left enabling the bird to lay its egg and the hatchling to develop and fly away, then this nest could be taken. Interestingly the young bird returned to the nesting area to eventually mate and build its own nest in the place it was born. This system had to be carefully monitored ensuring that nesting only took place twice per year, the result being that the bird population grew ensuring more nests as the years went by. Many bird nesting houses have now been constructed in Miri and other areas of Sarawak. Miri has around 30 bird nesting buildings. They are tall concrete structures with locked gates and holes in the walls to allow the birds to fly in. The buildings contain humidifiers to regulate the temperature and humidity so are quite expensive to build. The owners stick rubber nests on the beams and play bird chirping sounds using tape recorders to attract the birds. They utilise a pole with a mirror attached to check the nest for egg or hatchling. This system has greatly reduced illegal nesting in caves as it's a much safer system rather than climbing steep walls which does result in deaths from falling.

In Gunung Mulu National Park-many years after the above survey during the time I was based back in the UK, numerous groups of illegal nesters had been raiding the caves over a 10 year period. Revenue going into their pockets amounted to many millions of Malaysian ringgit but they had little to show for it except for hangovers. The unsustainable nesting caused a drastic reduction in the bird population numbers plus pollution of the caves and destruction of the cave fauna species. It was reported that hauls of 100 kg of nests had reduced to 5 kg. Their excuse for stealing nests every month was that if they did not go someone else would. Strangely no arrests were made. It can only be described as surprising that this was allowed to continue in a World Heritage Area.

The limestone of the Middle Baram is the same as Mulu National Park, the Melinau Limestone Formation of Late Eocene to Early Miocene age but here dipping at 10 to 30 degrees. The limestone of the 28 recorded isolated karst outcrops is only about 60 metres thick and rarely rises more than 70 to 80 metres above the surrounding Setap Shale. More outcrops are probably hidden by dense tropical rainforest cover. Clearly much of the karst is buried, as became evident during recent excavations for the Kejin Mini Hydro Power House near Anau, where solid, heavily fluted limestone bedrock was encountered less than one metre beneath the surface.

Again with the splendid aid of my Mulu Park guides we soon got down to some serious work on mapping. From north to south the Maloi area is surrounded by extensive oil-palm estates near the Maloi River. The full extent of the karst is currently unknown but, from preliminary investigations, many caves are known to exist. 42 entrances have been recorded, with only the one surveyed. This is Lubang Mering Jau Sing, at 528 metres in length and was mapped by a couple of the local community and me. The entrance is a large collapsed doline at the top of the hill, with a small guard house. The doline breaks through the shale overlay into the limestone and cave below. A steep descent leads to a gate and a classic phreatic passage, 8 metres wide by 10 metres high almost straight for 200 metres. A couple of high level leads could be seen but could not be entered without climbing equipment. The main tunnel appeared to be choked beyond a boulder pile but was not investigated.

The main way on is up a short climb to the south to a static sump. The sump is situated 38 metres below the entrance so is probably on the same level as the surface river, although no resurgences are known. To the east the passage continues to a junction leading to a climb up, overlooking a pit. The passage beyond almost certainly connects through to the chamber close to the entrance but was not surveyed as we had no spare rope. From the junction a further climb up and then down leads to a nice phreatic tube to the chamber near the entrance. This is a pleasant sporting round trip back to the entrance doline and would be suitable for adventure tourists at some time in the future.

Gua Payung was briefly looked at but was not mapped due to lack of time. This impressive cave consists of a 30 metre by 30 metre phreatic tunnel with fine cave formations, piercing a karst block and is reported as having thirteen entrances.

Forest Department Sarawak Licensing Unit records indicate that 38 caves have been licensed for nest collecting within the Maloi region at various times. All the caves contain *A. fuciphagus* colonies and are guarded by full-time employees of the licence holders, with a number of buildings constructed as guardhouses. The region is therefore reasonably well protected but there are so many more caves that require exploring and mapping.

The Batu Gading and surrounding area is under quarry licence. Because the several karst outcrops have largely been destroyed, along with the caves and any associated fauna, no detailed investigations could be carried out.

All the caves in Batu Basungai, Betok and Alat karst areas did contain small numbers of White-nest swiftlets but here again most of the caves have been destroyed by quarrying. Some of the licence holders eventually reached agreement with the quarry company regarding compensation.

Tamala karst area has a cave recorded on the available geological maps but attempts to relocate the cave have so far been unsuccessful.

The Sepayang karst was reported to contain over 15 cave entrances. We made extensive surface surveys and mapped three caves in this area: Lubang Urung, Lubang Pedan, and Lubang Ulang which all support colonies of *A. fuciphagus*. Other caves nearby are not yet surveyed but are known to contain small White-nest swiftlet colonies. The resurgence for the stream sinking at Lubang Ulang is unknown, but it is presumed to flow eastwards, towards the Sungai Sepayang, which marks the boundary of the Telang Usan Protected Forest.

A verbal agreement was reached with the quarry company that a 60 metre buffer zone would be left between the quarry face and the caves. Unfortunately, blasting within a few hundred metres of caves probably leads to roof falls due to seismic vibrations, as well as disturbing the swiftlet population. A minimum 500 metre buffer zone is required. It is also important to include in any proposed protection the catchment areas of the streams feeding into the caves, in line with international guidelines. Within this area a southwest–northeast-trending gorge is a fine example of a collapsed cave system.

Lubang Urung is situated at the base of a doline, approximately 500 metres to the north east of Lubang Pedan along a well-defined trail. From a guard house at the top of the doline a descent on bamboo ladders and a traverse over pinnacle karst leads to a 9 metre pitch. A lofty passage leads to a traverse over a hole in the floor to a calcite vein. A muddy ascent follows where I used an aluminium ladder for aid. Further muddy climbs and traverses follow to a 6 metre pitch down a calcite slope. A small inlet enters at the base of the pitch into a breakdown chamber where the gate can be found. Beyond the gate is the main nesting area where a 5 metre climb up reached a further pitch down of 6 metre. Following a low passage a climb upwards leads to the final sump and high aven. Again with the help of two of our Mulu guides, Robert Gani and Thambi Ipoi, we mapped the cave for 500 metres.

Lubang Pedan is situated approximately 650 metres due north of the Lubang Ulang guard house is in the valley. A well defined trail leads from the guard house through gardens and upwards skirting the edge of the valley to the Pedan guard house. The entrance doline is well fenced and a 5 metre pitch leads to the massive bomb proof reinforced concrete gate with double locks. A grill allows the birds to fly unimpeded through to their nesting areas. To the right a short passage with bats and well developed swirl holes in the roof, leads to a second entrance at the base of the doline. Through the gate a high passage passes a huge aven on the right. A wide and high guano-floored passage follows to a junction. Right leads to an inlet which can be followed until it becomes less then body size while left leads to the final calcite choked aven with a small inlet entering from the left. We mapped the cave for 340 metres, again only short in length but incredibly valuable as the cave is well protected with a sustainable nesting policy so the numbers of nests collected increases year by year.

The third cave in this area is Lubang Ulang. This is situated just 20 metres from the guard house in the valley. The cave acts as the main sink for the streams draining from the north and west but the resurgence is presently unknown. This is a fine complex cave not fully explored and mapped. The 10 metre wide main entrance leads to the streamway which sumps after a short distance. A passage to the right leads to a high guano-floored passage and to a massive area of breakdown terminating in a choke with a small inlet and sump. Daylight can be seen above from numerous high level entrances. The way to the upper levels involves a 12 metre slimy guan-covered climb up. At the top a passage shoots upwards to the west which was not surveyed. A further large inviting passage can be seen opposite at the other side of the passage below, involving a dangerous traverse across the 12 metre drop, again not surveyed but partly explored. The main way on is to the south to a four way junction. Right to the west is a fine phreatic tube with a vadose canyon in the floor. This leads to two small entrances in the cliff above the main entrance. Right to the south west leads via a sandy passage to a complex area of small passages again not fully explored. The main way on is to the south at the four-way junction. A climb up on the left leads to the massive breakdown area overlooking the main low level tunnel, with the high level entrances visible. Straight on along the south tunnel leads to a 4 metre pitch down through boulders to the streamway and sump. The low inlet passage to the south west was followed for approximately 30 metres but not mapped. Above the 4 metre pitch a large chamber can be entered with a climb to the left and a high level entrance. This climb arrives on a ledge overlooking the chamber below and exhibits fine underground pinnacle karst. From the chamber below, a difficult climb up reaches a large passage eventually leading to yet another high level entrance. This extensive and complicated cave is not fully explored or surveyed having at least twelve entrances; we only had time to map 400 metres.

Metarae karst area includes three separate limestone outcrops that had not been mapped or recorded in previous literature. Here the Melinau Limestone, resting on the Temala Member of the Kelalan Formation reaches barely 50m above the surrounding hills. Covered by primary rainforest the limestone outcrops contain sharp pinnacle-karst which is almost impossible to traverse across and few caves are known in this area. As the terrain is severe it has not been fully explored in detail. Its diverse flora is typical of limestone pinnacle-karst areas and no faunal studies have been undertaken but a wide range of mammals and birds has been observed. Though the region was proposed as an eco-tourism zone in later years, my development and management plan was never implemented owing to the untimely death of the land owner.

The region can be reached via a 3 kilometre, 45 minute walk through gardens and plantations, starting from the river bank opposite Long Laput, or alternatively along the Sepayang logging road.

The main cave in the region, Lubang Metarae is a classic sink to resurgence river cave averaging 15 to 30 metres in height and 12 metres in width with a guard house near the entrance. At the resurgence section a small collapsed lower entrance leads into the daylight-lit main tunnel, with a sealed higher entrance above and out of reach. The sculptured walls are impressive with horizontal solution grooves representing varying flow conditions. The main tunnel runs north east for 340 metres with a small stream occupying the left bank, until a collapse is reached and a sealed high level entrance. The cave continues beyond at a high level but could not be followed safely without aid. A route through the boulder pile reaches a sump after only 50 metres. The main way on is obviously above the boulder pile and must connect to a similar looking passage in the sink cave on the opposite side of the limestone block. The original cave development must have been associated with a very large river. This cave could be developed as an interesting tourist attraction. The cave is therefore divided into two sections by a natural barrier of collapse. It was not recommended to attempt to pass the collapse as the sink section of cave is under the stewardship of a separate Long Laput resident. The sink is a huge chamber 80 metres wide leading to a 50 metre wide section and a sump below the collapse. This was not mapped until 2012 during the planned eco tourism development study. The total length of this fine cave is 654 metres.

The boundary of the narrow northern limestone outcrop of Batu Metarae Ukat has a circumference of 1.23 kilometres and an area of 0.1 kilometres^2. Only one small cave is known, close to the western end of the block.

The middle outcrop, Batu Lubang Metarae, which contains the Metarae River Cave, has a circumference of 1.7 kilometres with an area of 0.185 kilometres^2.

Batu Paya Wan Uson southern limestone outcrop has no known caves. Its circumference is 2.23 kilometres, with an area of 0.2 kilometres^2.

The Kejin karst outcrop and its main resurgence cave is an impressive area of karst and forest with several highly productive, well-guarded bird-nest caves. Residents of Sungai Dua were highly enthusiastic regarding potential establishment of a protected zone. The extent of the karst proved difficult to define as much of it is buried beneath sandstone and shale. I was made welcome staying at the Sungai Dua Longhouse and the cave guard helped me with the mapping as I had not been able to recruit any Mulu park guides.

Lubang Tuking, is the cave visited by Spencer St John in 1862 and even today is still an important habitat of the White-nest swiftlet. The cave is well protected with a guard house and gate and with a sustainable nesting policy; the future of the swiftlet colony seems assured providing the area can be well protected. The area can be reached via a two kilometre walk from the Sungai Kejin, wrongly marked on the topographic maps as the Kejin Ok, into a nice valley with a well constructed guard house.

The floor of the entrance passage is entirely sandstone, passing a number of short lakes and cascades to the gate. The passage terminates at a pool with a number of high level continuations. The inlet stream originates from a low passage which obviously floods to the roof in wet conditions. The trend is generally up dip following the sandstone limestone contact. A passage on the left of the pool passes a junction to a chamber littered with sandstone and shale breakdown. The passage could be seen to continue at the top of a sandstone wall. From the junction the second passage leads to another sandstone wall with a small stream entering. It was reported that this connects to the passage seen from the breakdown chamber. The measured length was only 372 metres

Lubang Beruang was mapped for 464 metres and is situated up the ridge from the guard house in the valley with a small building guarding the entrance. A steep climb down leads to the gate and a chamber with a population of the mossy nest swiftlet and much guano. To the left is low and connects through to the main chamber. Right over a hole in the floor leads to the main chamber containing a 20 metre tall stalagmite in the centre with a small inlet of water above. The passage closes down beyond. Both Lubang Tucking and Beruang are important habitat sites for the White-nest swiftlet.

Lubang Anau, a cave I mapped for 460 metres and its entrance is situated below a guard house and farm in a large collapsed doline. This cave is the only one of its kind in Borneo being entirely developed in thin beds of shallow dipping calcareous sandstone. Containing large passages with domed ceilings and speleothems, the cave can be classified as unique for the fact that it does not collapse in on itself as the beds are little more than 20 to 30 cm in thickness. This rock sequence is the furthest south and must represent the fringe of the limestone deposition in the area. A climb down leads to the gate and the main tunnel, to the east another entrance has been blocked with cement to deter poachers, while west leads to the main route and a collapsed passage to the north. The main way on is down a slope to a stream inlet entering from Lubang Avang on the left. Again this entrance has been sealed off with a cement cap. The passage becomes smaller and passes an area of selenite crystals on the left wall. Further downstream the passage is small with much sediment on the walls indicating backing up of the streamway. This is hardly surprising as it was reported that another cement blockage was constructed just beyond the furthest point reached by the surveying team. This passage is reported to be connected to another extensive cave system, Lubang Usun Ngau Gia, or Bat Cave which I did not have time to investigate. A resurgence which is flooded to the roof was noted close to the hydro generating plant power house and could be the main resurgence for the stream. It can issue a large river during times of heavy rain but no sink is known to be associated with this system. It is suspected that the drainage may well not be allogenic as first suspected, although the water has been seen to discolour during heavy rain. It is possible that the drainage is derived entirely from percolation water flowing along the limestone, Eocene sandstone contact.

Beyond the guard house a canyon leads to a large arched cave containing burial remains,-this runs directly below the guard house but time prevented an investigation. On the journey back to the Baram River along the gravel road and just to the left of the main water pipes which drive the turbines, the main entrance of Bat Cave was pointed out. This is a large collapsed doline in sandstone, approximately 20 metres deep, with a number of large cave entrances leading off at the base. The cave is reported as being a few kilometres in length and is well worth further investigation as this could be the longest cave in the Middle Baram Karst Region.

It was considered important from a geological perspective that this area is protected as the caves represent the only such examples of part sandstone caves to be found in Borneo.

The Batu Salai karst appears to be small in extent but I suspect much of it is buried. Lubang Salai is just a small chamber and was recorded as a highly productive cave for white-nest harvesting but the harvest had declined dramatically over the years. One reason was that the walls were wet with a thin coating of damp clay which inhibits nest building. The suspected direct cause is the felling of a large tree above the entrance allowing percolation water to seep into the cave through the thinly bedded limestone bringing with it clay leached from the overlying soil. This is a classic example of the folly of removing vegetation cover from karst regions. The vegetation has now been allowed to re grow and the walls cleaned and dried out and a sustainable harvesting management plan is in operation. The cave is permanently guarded, leading to a remarkable increase in colony numbers. The swiftlets are now well protected so this cave can only be described as extremely valuable. No information about the Salai karst, just to the north could be located.

Sadly, although I recommended along with cave, surface maps and the report that this area should be constituted as a National Park, no action was taken for over 20 years. Just a few years ago two small areas have been notified as the Kejin Tugang National Park No 1 and 2, both close to the Tamala River but no caves are situated in the areas mentioned. On enquiries it did appear that the community suspected that if constituted as a National Park, they would lose their community land rights and privatisation of the park could take place as had happened with Gunung Mulu. Therefore at present none of the caves are legally protected.

On a direct line north east of the Mulu National Park karst area is an outcrop of limestone on the Limbang River. The only reference I could find was a cave used many years ago as a burial cave. On investigation it was found that the area was a limestone quarry so no further work for conservation was possible.

I had many great trips into Niah National Park with its famous 40,000 year old cave paintings and archaeological site. The trips were mainly investigation of the bird nesting activities and the possibility of nominating the National Park as a World Heritage Area. Sadly quarry operation has been prominent on the south side of the mountain for many years so it seemed doubtful for World Heritage status for as long as the quarry operation remained active.

Tanjong Datu National Park is situated at the western edge of Sarawak bordering Indonesian Kalimantan. It is only small in area but with primary forest and a wonderful coastline having been constituted in 1994. The only access into the park is by boat. It's a nesting area for turtles but the wild boars have a habit of digging up and eating the eggs. Fences have been erected in an attempt to keep the pigs at bay. A few caves in granite had been reported containing small numbers of White nest swiftlets so I was asked to go there with Braken and Chris from our National Parks and Wildlife Department.

Niah Cave entrance. Photo: David Gill

The sea was rough and as there is no jetty we had to jump off the boat into a rough sea and wade ashore. With my laptop balanced above my head this was not an easy option.

A small office had been constructed so we had somewhere to sleep and eat. The caves mapped were only small, no more than 50 metres in length but did contain a small swiftlet colony. One cave was complex in a jumble of fallen blocks of granite with a small stream. It appeared from reports that Indonesians paddled around the coast and stole the nests. With permanent staff stationed on site this would help to prevent the poaching.

After a few days of stormy weather it was time to leave as we had run out of food, the problem was the boat could not make it due to the very rough sea. All we had left to eat was rice so we spent days searching the beach for snails. After boiling the snails a fish hook was required to get the meat out of the very tiny shell. It was a case of 'what is for breakfast today? Snails again, lunch snails, dinner snails, if you don't like them starve'. After a few days of snails the weather cleared and we could escape on the boat.

When guiding high-ranking government officials in Gunung Mulu I was often asked why the caves had English names rather than Malay. I tried explaining that they all had Malay names but due to the advent of tourism the English names were preferable. It was also pointed out that all the caves had been explored, named and mapped by British cave explorers during British expeditions. The answer was not liked and questions were asked on why the caves could not be explored by Malaysians. The formation of the Karst Management Unit was an attempt to solve the problem with the undertaking that all cave expeditions would be classified as joint ventures. It was up to me to train National Park staff in caving techniques including rope work and mapping with the hope that they would join in the fun. The expeditions always involved park guides as this was considered as good training but many years later when I was back in England it came to pass that their main interest was bird nesting so taking along cave guides was eventually dropped by the expeditions.

The best place to start training National Park staff was in the caves near Kuching at Bau. We had two caves as Nature Reserves, Fairy Cave and Wind Cave. I bolted a cliff face at Wind Cave and hung a series of ropes for teaching single rope techniques; we had mapping and rescue practice, all useful for members of the staff. The most able was Rambli Ahmad whom I had worked with at Gunung Mulu National Park, we became good friends and his family often stayed with us in Miri. In later years after my return to the UK, Rambli became the only member of staff that could be considered as being interested in loving caves and took part in all the Gunung Mulu cave expeditions. On the official formation of the

Mulu Caves Project he became one of the trustees and quite rightly so.

Training staff, Bau Caves, Kuching. Second from the left, Rambli Ahmad, Oswald Braken Tisen, author, Sapuan Hj Ahmad, Victor Luna Amin.
Photo: Unknown

The problem with the Bau limestone is that much of it is composed of isolated towers with licences issued for quarrying. The caves are numerous with many recorded and mapped by G E Wilford and published in his splendid book. Many of the caves are controlled by local community families engaged in bird nesting. I managed to map a few but realised that this was a mammoth task for a two man team and required full-blown expeditions in order to achieve anything of significance. Many of the caves are complex and extensive.

There are also a few old gold mines around which are of historical interest. As a tourist destination the area has great potential with magnificent tropical karst scenery and many caves that could be developed as tourist attractions. One area has now at last been constituted as a National Park at Dered Krian but when a quarry company began its operation many people complained as the area under the law is supposed to be totally protected. The quarrying was halted.

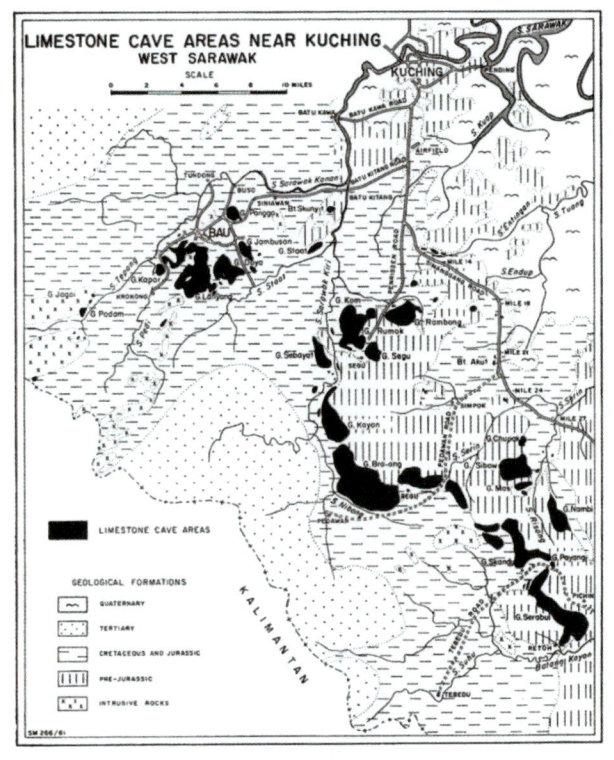

Wilford's map of the Bau karst area

Gunung Mulu National Park Pinnacles. Photo: David Gill

4
Gunung Api Connection Expeditions, 1990 to 1994

'This is all you have. This is not a dry run. This is your life. If you want to fritter it away with your fears, then you will fritter it away, but you won't get it back later.'

Laura Schlessinger

In 1990 Alan Weight organised an expedition to Mulu along with his wife Becky, Tony Bennett, Steph Gough, Rick Halliwell, Nick Thompson and Paul Norman. Simon Lagang and Wilson Bala two park guides had discovered a cave on the banks of the Melinau River provisionally called Simon's Cave. I later re-named it to Racer Cave as the snakes coil up at a constriction and capture passing birds or bats. It was park policy not to name caves in the park after people. The cave had been surveyed for 900 metres during the Mulu Caves Expedition in 1989 but at the furthest point a lead could be seen up in the roof. The following year Simon and Wilson, two of our park guides had managed to levitate up the climb to find a considerable amount of passage beyond. They were probably looking for swiftlet nests. Alan and his team mapped the cave to around 3.5 kilometres in length and the interesting point was that it passed above Lagang's Cave and was also close to Wind Cave. A connection seemed to be a distinct possibility.

An attempt was made to connect Drunken Forest Cave to the Clearwater 5 cave river upstream. The connection was there but only 30 cm wide so a tape was passed through to prove the connection between the two caves. Also of note was the discovery of Clearwater 6.

Alan and Becky returned in 1991 to clear up a few leads helped by park guides. This was highly successful mapping more in Racer Cave, exploring Palm Cave and mapping Fern Rock Cave with a length of 845 metres. Palm Cave was finally mapped Matt's expedition in 1991.

Alan's follow up expedition was in 1994 as the Gua di Gunung Api Expedition. With Alan was his wife Becky, Simon Ashley, Pat and Ric Halliwell, Paul Norman and Simon Parker.

The Shower Head, Racer Cave. Photo: Robbie Shone

I was also around with two caving friends Vaughan Thomas and his wife Lynn Roberts as they often popped into Mulu on a regular basis as Vaughan worked off shore in the oil fields of Brunei. To make life easier Vaughan had also purchased a boat and outboard engine which finished up with me when he was transferred. Accompanied by ten park guides on and off a few caves had been explored and mapped with Alan's team concentrating on numerous potential leads.

One of the main problems with expeditions to explore caves in Mulu was the compilation of a final report including cave maps and any science that had been conducted. This is a difficult and time consuming task but unless completed and submitted to the national park it was impossible to keep track of all the science and exploration. The same applied to photographs. The Forest Department had given permission and permits for two expeditions, one from Indonesia and one from Korea. Both claimed to have explored many kilometres of caves but no report or map was ever received. I even wrote letters to the leaders but never received a reply. The good point to note about all British expeditions was that the reports, maps, scientific results and photographs always arrived so we knew exactly what had been achieved with possible leads remaining to be explored.

Alan's first major breakthrough was in Leopard Cave from Septic Scenic Series where a hole was found leading after a 10 metre pitch to my marked survey station in the Blackrock to Clearwater connection passage. This was an easier route into Blackrock and enabled the team to investigate a number of potential leads but none led into major extensions.

The Northern Entrance to Clearwater was located with difficulty as it is just a tiny hole in boulders. A few potential leads were followed but again did not lead into a huge extension of the known cave. The connection from Leopard Cave to Clearwater brought the total length of the Clearwater, Blackrock system to 107 kilometres.

Wind Cave had a number of areas that required further investigation. The upstream end of the main river passage was close to Racer Cave but nothing of note was found that could lead to a possible connection. The Seven Pillars of Wisdom was another matter-being complex with many possible openings that required exploration but again no major extensions were discovered.

Peter's Cave was mapped in the isolated southern block of limestone accessible via the Deer Cave plank walk. This is a five metre wide phreatic passage with hoof prints of wild boar in the clay floor and there is also a partly explored high level series. There has been much confusion over this cave as it has been mapped three times, all by different groups. Besides Alan's map it was partly re-mapped in 2008 by three members of the Malaysian Nature Society-with three park guides who named it Wild Boar Cave. It was again partly mapped by the Mulu Caves Project in 2010. It is presently only a few hundred metres in length but it is suspected there is much more to explore. Wild Boar Cave is the preferred title.

Stone Horse Cave was mapped by me for 2.2 kilometres in 1991 and is a large 20 to 30 metre wide phreatic passage with imposing formations and a few high level entrances. It is situated 40 metres above the Deer Cave plank walk so is easily accessible. I spent many days measuring the cave for a walkway and electrical lighting but no decision was ever received on the proposal.

As an adventure cave for tourists, I fixed a few bolt anchors to the walls and fixed a rope to aid the traverse across the deep hole in the floor. Sadly the ropes disappeared, presumably stolen by poaching bird nesters. Having caving friends around much work was accomplished in this area with the discovery of Frog Cave mapped for over one kilometre, this is a complex cave and not recommended for adventure caving. Eventually a 25 metre pitch led down into Stone Horse.

Fern Rock Cave is situated to the north of Stone Horse in the cliff above the Melinau Paku River and had been surveyed during Alan's 1991 expedition for 845 metres. More entrances were mapped and a connection made to Stone Horse bringing the length to 1.66 kilometres.

Paris Cave is high above the plank walk over the usually horrendous karst terrain of thick vegetation, pinnacles and crumbling rock. It can be easily reached via Entrance 4 in Stone Horse. The main entrance is 50 metres wide and 25 metres high with marks in the mud floor of wild boar. The passage narrows and at 30 to 50 metres high enters a chamber with a 25 metres pitch into Fern Rock Cave. This system is now 6.8 kilometres in length with probably more passages to find.

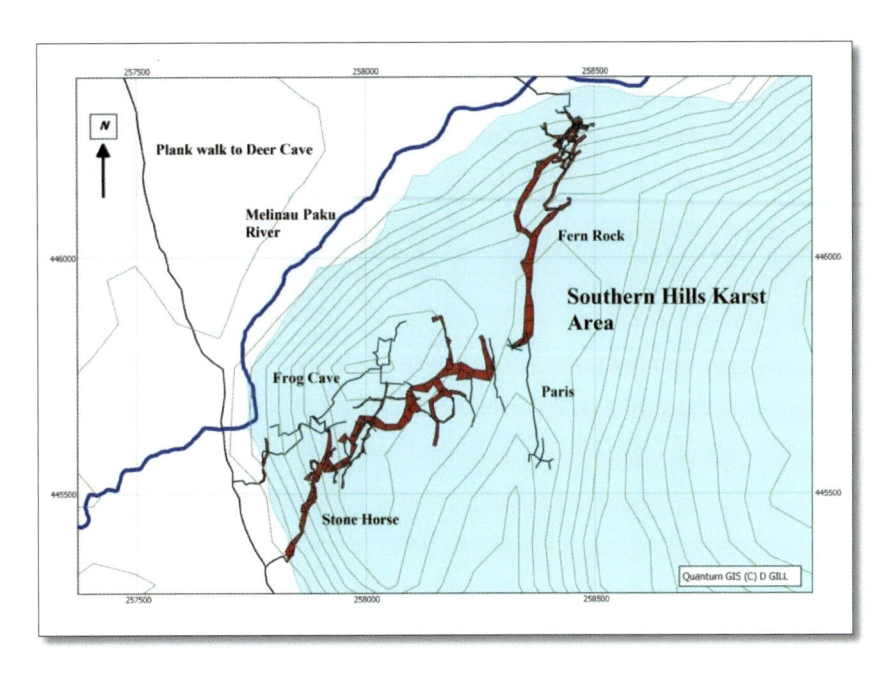

Stone Horse Cave System

Hidden Valley, Gunung Mulu National Park. Photos': Rambli Ahmad

Camp 5 in the Melinau Gorge was also visited and a search for caves conducted but nothing substantial was found.

Alan's expeditions were highly successful, providing yet more knowledge of the ramifications of the Mulu world-class caves systems.

5
Reconnaissance to Hidden Valley 1993

'Let your dreams be bigger than your fears, your actions louder than your words, and your faith stronger than your feelings.'

Unknown

The best way of describing Hidden Valley is remote but at the same time spectacular. It is situated on the east side of Gunung Api reached via a three hour walk along a well-defined track from park headquarters crossing the Melinau Paku River to Camp 1. There is no bridge across this river, so in times of heavy rain, it's impassable. Onwards from Camp 1 which is a small timber hut, an uphill slog with no definite track leads to a north east ridge at an altitude of 950 metres. This passes a major tributary to the Melinau Paku with a thundering great waterfall close by, a nice place to stop for a sandwich. From the ridge which eventually leads up to the summit it's a very steep descent down into Hidden Valley. The walk in was around 14 to 16 hours.

Two major caves had been explored and mapped during the Royal Geographic Society expedition in 1977/78 but Hidden Valley had not been visited since. With the connection of Blackrock Cave to Clearwater in 1991, it was considered time to look at other possibilities. It was thought that Hidden Valley was a prime target but a reconnaissance was required as a first step considering the difficulties of access. We had a strong team with Matt Kirby, Tim Allen, John Palmer, Vaughan Thomas, Lynn Robinson and me. With four prominent park guides and four Penan porters it was a large crowd.

The Penan were in this case considered to be essential as they know the park like the back of their hand and seem to be able to pick the best route without getting lost.

In Mulu it's easy to get totally lost as a park guide and I found out on a number of occasions. The first was an inspection trip to Drunken Forest Cave where we slept the night in the entrance. On the way back to headquarters we lost our trail-which was only marked with tiny cut branches every few steps of the way. We wandered around in circles until we finally realised we were back at the same place we had left one hour ago. Yet another was at Camp 5 in the Melinau Gorge with one of our Forest Guards and a guide after a day cave entrance searching. We left it late and it went dark on us as we stumbled about in the forest getting absolutely nowhere. Finally I decided just to follow the cliff line which led us back to Camp 5.

We set up a small temporary shelter in the valley away from any potential flood water with the 500 metre high cliffs either side of us. The stream in normal flow conditions sinks just beyond Prediction Cave but overflows during heavy rain to sink again at the end of the valley. From this point the pinnacle karst commences being desperate to navigate with holes everywhere covered in fallen trees, vegetation and loose boulders all in a dense jungle. We were able to utilise the dry stream bed as a roadway to begin searching the dolines marking the westward route into the karst. It was also possible to view the adjacent cliffs and three entrances were noted 50 metres and more above the valley floor. Two major dolines were investigated and the valley could be seen trending to the south towards the Melinau Paku Valley which appeared to contain many deep dolines well worth further work on a future expedition. To the west was similar country heading toward the Clearwater Cave System, indicating the original route of the drainage. One of the dolines contained a pit with a strong draught which looked promising but we were not carrying any ropes so could not descend. Another two cave entrances were seen high up on the north wall of the valley, again requiring substantial climbing equipment to reach them. Over the next few days yet more high level dark-looking cave entrances were noticed, obvious projects for any forthcoming expedition. In total the team had found 17 potential cave entrances with the valley trending south towards the Melinau Paku Valley and also continuing to the west.

We had a VHF receiver and transmitter on the summit of Gunung Buda so as a backup I had taken along a hand held radio receiver transmitter and had managed to contact our headquarters. I received a surprise message to return as soon as possible as my boss Braken was on his way and wanted me to join him on a trek to the summit. This was to inspect the trail, Camp 1 and Camp 4, the final camp on the mountain situated often above the clouds. I set off from Hidden Valley with a Penan porter and after a very long and tiring trek arrived at Camp 1. The following day we headed back to headquarters where I arrived bleeding heavily from numerous leech bites. The problem with leeches is that you just do not notice them and they worm their way into your socks or embarrassingly your private parts. As they inject an anticoagulant and analgesic you cannot feel them sucking your blood and the hole emits blood for a long time afterwards. Finding your boots full of blood can be very disconcerting.

After one night of sleep in bed I was off for the four day journey to the summit and back. This is a 24 kilometre trek to a height of 2,377 metres, a long and hard climb through jungle. We spent one night at Camp 1 and one night at a very cold Camp 4, then on to the summit. The journey back we spent one night at Camp 1 before returning to our headquarters. It was established that the trail was in reasonably good condition but required a Camp 3 which was up to me to go and build.

During the years at Mulu I continued to have much fun as working there could only be described as a privilege. Our quarters became something of a holiday resort for visiting British cavers as visitors arrived frequently. Liz Price stayed for a while, a prominent caver living in Kuala Lumpur and a member of the Malaysian Nature Society. Steve and Kath Jones turned up and we completed the through trip from Wind Cave to Clearwater without getting lost. Even team members of my Untamed River Expedition to New Britain visited, Dave Sims, Dave Arveschoug and Ken Kelly. A constant visitor was Dave Clucas who popped in usually after a business trip to Singapore. Dave loved Sarawak so much he eventually married Betty's Aunty, bought a house and built another one near Limbang town. It was always a pleasure to see Matt Kirby and his wife Liz who was bitten by a snake on the plankwalk outside my house, not fatally. We even got lost together searching for Lagangs Cave but found it eventually without crying.

It was also very easy to get completely lost in the giant Mulu caves so adventure caving tourists needed to be guided by a competent guide who could find his way around. Two Danish ladies decided to do the through trip from Wind Cave to Clearwater with a guide. By midnight they had not returned to park headquarters so I sent out two search parties. We found them in Clearwater none the worse for wear. The guide needed educating on route finding. One other tourist became totally lost again with a guide in Lagang's Cave spending all night wandering around looking for the entrance.

The Hidden Valley helicopter landing zone. Photo: Matt Kirby

At this time we all used carbide lights and carried spare carbide with the explicit instruction to bring the old used carbide out which resembled white powder. No carbide no light, no light and you are lost. Just a few years ago a tourist wandered off to Camp 1 without a guide as is prescribed by the management. He became lost and was missing for many days before a search party finally found him still alive.

6
Hidden Valley expedition 1996

'Courage is the most important of all the virtues because without courage, you can't practice any other virtue consistently.'

Maya Angelou

After the 1993 reconnaissance expedition a return for a full scale expedition to Hidden Valley was inevitable. The main problem was how to get team members, food and equipment into this inhospitable place. As a joint venture with the National Parks Department I was able to hire a helicopter-problem solved.

Matt Kirby was leading the expedition and had a great reliable team including Tim Allen, Richard Chambers, Mark Wright, Pete Boyes, Dr Pete Smart our scientist, Mick Nunwick and Nick Jones. Besides myself and Rambli Ahmad I had my National Parks team of three guides, Syria Lejau, Lawai Kumpang and Christopher Jantan, also our highly dependable Penan, Tamajarau. Betty's brother Charlie Usang was our camp manager and cook. We finished up unexpectedly with another two, Louise Korsegard and her sister Kristina. Mention had been made at my office that Louise wanted to join the expedition. I advised against it as Hidden Valley is not the place to take anyone without vast experience of expedition caving into a tropical jungle which was a remote and potentially hazardous place. It appears she obtained a letter from the Director of Forests saying that she could join the expedition with permission from the leader who was Matt but Matt knew nothing about it. Both ladies were regular visitors to Mulu enjoying the adventure caving so were hardy in nature and their father was a highly respected advisor to the Forest Department Sarawak. It was noticeable that cavers involved in Mulu became hooked on exploring the huge and complex cave systems so it was understandable that they would want to join an expedition at some time.

The surprise came to the team at the International Airport as they were leaving for Malaysia as Louise was there at the same time and asked them if they were cavers. You can't miss them anywhere as a group of cavers sticks out like a sore thumb. When they told her where they were going she pronounced that she was going as well, surprise, surprise. For the sake of not causing a diplomatic incident things were smoothed over and they both came along without any problems occurring.

I decided to send off the guides and Charlie early for the long walk in armed with parangs and a few Penan porters. They were to build a camp and a helicopter landing area. The majority of the team decided to walk in while a few of us organised the equipment and food supplies for the helicopter supply. The pilot had some misgivings trying to land on a very small landing zone surrounded by jungle and with 500 metre high cliffs surrounding the site but with some highly skilled flying he succeeded.

As for communication the VHF set on the summit was out of action so we procured two Single Side Band long wave radio receivers and transmitters. This involved setting up ours with a long horizontal aerial up on the ridge. This was a long climb from our base camp but preferable than nothing. The second set was situated in the Mulu Canteen. This meant we could contact the headquarters and the helicopter company in case of an emergency. It is what I have come to refer to on expeditions as a contingency plan.

We were now at last set up to begin our exploration of the potential leads we had noted on our reconnaissance expedition in 1993. Two of the high level entrances were climbed by Tim Allen and Mick Nunwick. The first entrance at 150 metres above the valley floor proved to be an alcove while the second at 70 metres above again proved to be disappointing with no enterable passage. A few other high level entrances were spotted-mostly in the northern wall but were considered as major climbing projects so were not attempted.

The stream that flows into Hidden Valley sank below the surface through rocks and gravel but at times of heavy rain-which seemed to happen every-afternoon overflowed the sink to disappear once more further along the valley. The upstream sink had been dye traced during the early Mulu expeditions to the river in Nasib Bagus, Good Luck Cave being the site of Sarawak Chamber. Beyond Sarawak Chamber the river passage sumped but was heading directly towards Hidden Valley. We found out the problem of heavy rain during the expedition when the kitchen was washed away.

The pit we had found during the reconnaissance expedition in the 2nd doline passed another pit guarded by a small Yellow Pit Viper snake, no one wanted to descend it even though it was draughting. Three weeks later it was still there waiting for a bat or caver.

We descended the main pitch with the aid of a rope to a ledge. From the ledge two ways on led to a couple of blocked passages and a large passage to a further doline entrance with a further passage blocked with sediment. From the ledge an 18 metre deep pitch led to a further 11 metre drop leading down a slope to a small streamway.

Upstream through a sediment fill was blocked after 150 metres with the stream issuing from a high aven in the roof being just percolation water. A short distance upstream from the pitch a tube led upwards to a large tunnel. To the right led back to the stream passage at roof level but the main tunnel continued to a calcite blockage with a low passage leading off through boulders. This terminated in an upward climb with loose rock everywhere. Sometime later the surface pitch with the Yellow Pit Viper was descended by our park guides for 20 metres followed by a further 10 metre pitch. At the base they found one of my survey markers. From the base of the 18 metre pitch further passage was discovered but blocked with sediment. I had a very enjoyable trip with our scientist Peter Smart collecting and noting bits of gravel. We gave it the name Arch Cave and it was mapped for 980 metres and 106 metres in depth.

Christopher Jantan and I spent a day thrashing through this horrendous karst along the north wall attempting not to fall down hidden holes in the limestone. We did not find any open caves except a few rock shelters with paw marks of a large wild cat, possible a Clouded Leopard. On the way back to regain our track that was now becoming well trodden we skirted around a doline where the dense vegetation prevented anyone from seeing more than a few feet. We noticed swiftlets flying through the trees which indicated a cave somewhere. Climbing down into the doline we found a large cave entrance heading, as with the majority of Mulu caves, at 30 degrees along the strike of the bedding. Footprints were noticeable indicating that some members of the reconnaissance expedition had entered but for some unknown reason had not recorded its existence. We decided to hand over the exploration of this cave to our park guide team so the following day I set off with Syria, Lawai, Rambli, Louise and Kristina.

Damocles Cave. Photo: Matt Kirby

We mapped to a massive boulder choke but a draughting small passage led down a few rope pitches to a stream inlet from a great unstable boulder pile. From the survey this appears to be directly below the final overflow sink in Hidden Valley. The stream disappeared down a rift passage with smooth water-worn walls.

The following day, armed with rope and bolting kit, the guides rigged the pitch down to a streamway, 39 metres below. This was followed for approximately 100 metres heading towards Cobra Cave in a rift passage which obviously flooded to the roof during the rain-which started around 3 pm every day. I had no choice but to call a halt as I did not want to lose any guides. We called it Damocles Cave and mapped it for 245 metres with a depth of 79 metres.

No Name cave was interesting due to its location and size. This was to the south west above the 1st doline but terminated in a huge choke. Looking across the valley on the north wall other cave entrances were spotted indicating that this cave had been truncated in the distant past by incision of Hidden Valley.

Matt had a good search in Prediction Cave without finding any possibilities as this 100 metre wide cave is full almost to the roof with sediment. It was estimated that the passage height must be in the region of 60 metres.

Above Prediction four cave entrances could be seen, the first one at 40 metres above was bolt climbed by Mark Wright and Tim Allen over a two day period. Sadly although the cave was 30 metres high it went nowhere. The other entrances were not attempted.

Perseverance Cave was found by Pete Smart and Matt 30 metres above the base of the 3rd doline. It was used as a bivouac site as there was a nice soft floor at the entrance and a water supply. Unfortunately it was inhabited by Moon Rats so did tend to disturb sleeping cavers. A 7 metre wide passage led via a small stream to a junction. The main route led to a second entrance. From the first entrance a passage on the left follows the cliff line and emerges in another large entrance with a small stream which can be followed for 100 metres to connect with passages from the first entrance. The main route was a tall canyon entered at a junction terminating at a calcite blockage with what appeared to be a small sump below. The main passage could be seen high up above the calcite blockage but was not reached. The cave was mapped for 953 metres.

The big breakthrough came with the discovery of Bridge Cave situated above the 3rd doline opposite Perseverance Cave. This was not an easy place to get through as heading in a south westerly direction towards the Melinau Paku Valley the terrain is as severe as ever with pinnacles, crumbling rock and thick vegetation.

The entrance has a narrow bridge across but using tree roots a descent was made to a pitch where we fixed a hand line. Beyond a passage on the right was later used as a bivouac site, the only drawback being it was used by other mammals and birds. Straight on was choked but a strong draught was apparent from a passage on the left which had a hollow calcite floor. I mapped onwards to a total choke but returning found a low passage where the draught originated and popped out in a huge white walled passage, White Cobra South.

The scale is impressive with a large scree slope leading upwards to an extensive series of high level passages. The passage follows the strike of the beds heading south towards unknown ground. After 300 metres the passage reduces in size with scalloped walls but a climb up to the left enters a complex series of boulder-floored passages and another entrance which overlooks the 4th Doline. Promenade Passage leads up dip from White Cobra into a passage parallel and 40 metres above. This area is far too complex to describe but of significance was a pit with the continuing passage seen beyond up in the roof, this is the most southerly part of the cave and represents the most significant lead but a great deal of difficult climbing would be required.

Bridge Cave. Photo: Matt Kirby

The Abyss is worthy of note as this is an enormous pit 30 metres in diameter. An attempt to descend proved far too dangerous with loose flakes threatening the caver. A Mars Bar wrapper was dropped down this 95 metre pit with the hope of finding it.

Twin Waterfalls Chamber was again a ramification of passages with a small entrance overlooking the doline containing Arch Cave. Tomb Raider Series was yet another complex maze of old cave passage typical of the Mulu caves. Over 6 kilometres was successfully mapped in Bridge Cave, a substantial find filling in a great blank area on the map.

Cloud Cave was found after a great deal of difficult terrain was traversed by Tim, Mick, Lawai and Chris. It is situated beyond Bridge Cave on the north east side of the 4th doline and gets its name from the cloud within the entrance. The entrance is massive and a boulder slope leads to a 37 metre deep pitch. Below the passage measures 80 metres high and 45 metres wide, Kamikaze Highway. Upstream a huge boulder slope leads to the base of the Abyss found in Bridge Cave 95 metres above. A 10 metre pitch behind a forest of large stalagmites led to a boulder choke with no way on. Crab Inlet was found hidden behind sediment banks and was mapped for 350 metres in a large passage totally blocked with boulders.

Bridge Cave. Photo: Matt Kirby

Downstream a scree slope descended for several hundred metres to a 10 metre pitch where beyond an immature stream passage was entered. Wisely, exploration was abandoned due to the imminent possibility of extreme flooding which happens often in this part of the world. Just above the pitch a traverse reaches a side passage where it connected to the furthest reaches explored in Cobra Cave, the sought connection had been made!

Mark Wright, Syria and I camped in the entrance for one night. It was amusing as the millions of swiftlets kept flying into us in the evening as we were not supposed to be there. At times we had to fish them out of the soup. This cave has the highest number of swiftlets of any Mulu cave, possibly several million as the morning exodus continued for many hours. Thankfully they nest in the roof 80 metres above so should be out of reach to the bird nesting poachers. The Last Tango in Mulu was a side lead at the entrance with a descending passage to a 30 metre pitch. This route led for 500 metres to a 5 metre pitch ending in calcite boulders. Above a void was seen but would require a very serious climb to enter. Below the pitch a massive rift passage full of boulders was encountered, when rocks were thrown down it took 10 seconds to reach a presumed floor level. This was not descended but is assumed to connect with Cobra Cave far below. Cloud Cave was mapped for 2,192 metres with a depth of 220 metres.

The Wonder Cave accident was just one of those very unfortunate incidents that could happen to anyone. Mick Nunwick and Nick Jones had set off to see if it was possible to push beyond the furthest point reached way back in 1978. At a traverse there was an old piece of polypropylene rope from the 1978 expedition. Mick passed it without incident but Nick must have placed some weight on the rope and it snapped sending Nick bouncing down to the bottom. Mick Nunwick managed to reach the base of the drop finding Nick in a bad way. His helmet and light had been damaged and he had a bad cut to his head. Luckily with no broken bones Mick managed to get him out to the surface, an amazing feat on his own. Dr Peter Smart stitched up his head and as it was late at night there was nothing anyone could do until the morning. We sent Tamajarau off early morning with a note back to base camp as there was no way that Nick could walk out. A helicopter air lift was the only means of evacuation. We set up the Single Side Band transmitter at the top of the ridge and managed to contact Bako National Park but not Mulu. The chances of obtaining a helicopter seemed slim. Tamajarau amazingly reached base camp in 3 hours 50 minutes from Hidden Valley; he must have been running all the way. The manager of the canteen Mr Ting paid cash for the helicopter and it landed safely with Ting in it as the pilot had no idea where Hidden Valley was. Nick was evacuated to hospital accompanied by Mick Nunwick. He had a very uncomfortable journey back to the UK but it was not long before he was back in Mulu.

The expedition had considerable success with over 10.5 kilometres of caves mapped. The Cobra, Cloud and Bridge Cave System were now at 15.5 kilometres in length with many leads left to explore.

The team in Hidden Valley. Left to right, Matt Kirby, Dave Gill, Tamajarau, Mark Wright, Peter Smart, Tim Allen, Syria Lajau, Richard Chambers, Charlie Usang, Pete Boyes, Rambli Ahmad, Mick Nunwick, Lawai Kumpang and Christopher Jantan.
Photo: Unknown

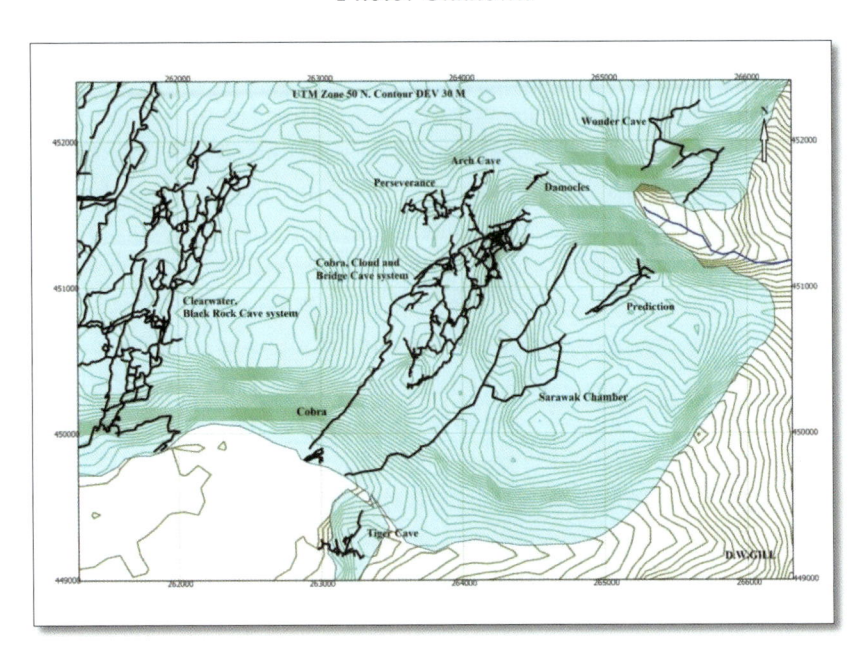

The caves of Hidden Valley, 1998

7
Hidden Valley 1998 expedition

'You are not a pawn in the chess game of life; you are the mover of the pieces.'

White Eagle

It was not long before Matt decided to come back to Mulu as he just can't keep away, this time with Tim Allen, Mark Wright, Nick Jones, Rupert Skorupka, Peter O'Neill, Martin Holroyd, Peter Hall and Lee Cartledge. I had a small team including Syria, Robert Gani, Tamajarau and Charlie our cook. Matt decided to set up a base camp near Nasib Bagus Cave, a wise decision as we had a good water supply and it looked possible to reach the 4th doline by climbing up the scree slope and traversing north along horrendous karst terrain with pinnacles and lots of holes to fall down. There was always the alternative of access to Cloud Cave via Cobra Cave via the connection found on the previous expedition, a long way but achievable. My major interest was to map Tiger Cave as this had been found during the 1978 expedition but not mapped. The camp was set up being one and a half hours trek from park headquarters along the trail to Nasib Bagus. This was a very enjoyable camp as we had a supply of scotch and European visitors arrived staying the night and carrying more scotch. The morning wakeup call was the melodic songs of the jungle birds.

Attempts were made by the hard men to enter the western valley from Cloud Cave but the karst was so severe they spent 10 hours to gain no more than 50 metres. Helicopter Cave was entered again via desperate pinnacle karst to an 8 metre climb up but the cave was totally choked after 30 metres.

A bivouac was established in Perseverance Cave and the Highest Cave in Mulu was reached by Tim Allen, Peter O'Neill and Robert Gani. This is at an altitude of 580 metres. The south side of the doline was descended to a large rift entrance. To the east the rift led to a daylight entrance with massive calcite formations. From the entrance a 20 metre high rift led to a 47 metre pitch but half way down the rift was seen to continue heading west but was not reached, this remains unexplored. A further 10 metre pitch led to a large chamber 100 metres across and a further 8 metre pitch led to a few descending passages all blocked with no draught present. The Highest Cave in Mulu was mapped for 280 metres in length.

Western Front Cave situated high above the 3rd doline terminated in a large chamber with a crawl leading to a higher entrance.

From the helicopter in 1996 we could clearly see the dolines stretching away to the west towards Clearwater and unknown ground. Dark patches could be seen in the dolines through the heavy tree cover but it became obvious that this terrain was impenetrable to us mere humans.

Mark Wright, Syria, Robert and I decided to undertake a detailed examination of the final lake in Cobra Cave. About one kilometre in a short pitch down was rigged with a rope and beyond we discovered a bird nesting poacher's camp. It seemed probable they were in the cave hiding somewhere. On arriving at the huge final chamber we descended down to the lake with no signs of the poachers. They were probably hidden with their lights out. The lake did sump to the roof in wet weather but the weather at this time was good with little rain. It was decided that this would be the best way to gain access to Cloud Cave and Hidden Valley rather than the highly dangerous trek across the karst. On returning, the rope down the pitch had disappeared, obviously stolen by the poachers. Was this an attempt at assassination? Robert bravely free climbed the pitch, rushed to the surface and camp to bring back a rope. We exited safely and although I reported the theft to the police, of course nothing could be done. Some months later I accused the suspect but of course he denied it.

The calcite blockage in Perseverance Cave was passed via a roof traverse and a short pitch beyond leading to a 20 metre wide passage. 180 metres along another entrance was found arriving back out to the 3rd doline. 600 metres was mapped bringing the total length of Perseverance to 1,759 metres.

A further 3.2 kilometres of cave was discovered and mapped in Bridge Cave bringing the total length to 9,896 metres but little was found in Cloud Cave. The total vertical range of the Cobra, Bridge and Cloud Cave System came to 450 metres

A cave entrance was seen from the Melinau Paku Valley but it was

150 metres above and attempts to reach it failed but the Invader Streamway, which is the resurgence from Cobra Cave, revealed 618 metres of very wet passage to a sump. This cave is without doubt highly flood-prone and should be avoided if it's raining.

Bridge Cave. Photo: Matt Kirby

Tiger Water and Tiger Back Caves were explored and mapped by Syria, Peter Hall and me for a distance of 1,582 metres; this is situated in a block of limestone called Batu Nigel. The limestone block is connected to the main Api Mountain by a pinnacle ridge. So there was no mix up with names we called it Tiger Water to distinguish it from Tiger Cave in Benarat Mountain which can be seen from Camp 5 in the Melinau Gorge. The stream sink is choked with debris but a large dry entrance close by descends to a choked downstream sump. Upstream a superb sculptured passage terminates at what appeared to be another small entrance. South-east leads to the main stream junction, upstream leading to a sump while to the left led to a nice waterfall from a high aven. Both the sump and waterfall are related to the main choked sink close to the entrance. From the main stream junction a nice streamway with many fish leads downstream in a high passage to a large cobbled chamber. A number of passages were choked and the main stream sumped but a flood by pass was followed which obviously sumps in wet weather. A tube by passes the passage with little airspace to a large cobbled floor passage to the resurgence entrance. Following the surface river arrives back at our base camp after a short distance.

Tiger Back Cave can be seen from Cobra Cave as an 80 metre wide entrance 60 metres above the resurgence. This was partially mapped and a few other shafts close by descended dropping into Tiger Water Cave. The cave was reached via a sandy climb to a balcony with this large cave entrance guarded by huge stalagmites. A large passage is choked being close to the large high level entrance above Tiger Water Cave on the opposite side of Batu Nigel. A number of possible leads in this area were noted but none have been explored or mapped to date.

Exploration of Clearwater V and V1 River was on the cards as the El Nino year had brought to Mulu stable weather with little rain. As water levels were low this was an ideal time to push beyond the previous limits, sadly I was too busy with work commitments but Matt's team decided to attempt to go beyond. It's a difficult area to reach as there are only two ways in, one via Armistice South discovered and mapped in 1989 and the second via Armistice North and a 43 metre pitch into Clearwater River V. It's a long way from home with no surface communication. In normal conditions the original route involved a one mile swim and wade in order to reach the upstream sump and Water Gate which was a bypass to the sump leading to Clearwater V1. Alan's expedition on returning to the sump found this bypass flooded which prevented exploration. During Matt's 1989 expedition the upstream sump was not reached due to high water conditions.

This area can only be described as dangerous as any heavy rain produces a rise in water levels which could prove disastrous for any cave explores in the river at the time. The original route was used by the team and low water enabled a pleasant stroll upstream to the sump and Water Gate. Clearwater V1 was only 50 metres long leading upstream to a large deep pool but beyond the pool a slippery climb led to a hole down and back into the streamway via an 8 metre pitch into Clearwater V11.

Downstream was a sump but upstream in a passage 10 metres wide and 20 metres high was mapped for 500 metres to another large deep and clear pool, yet another sump. There was only 980 metres separating this sump to the sump in Black Magic River which I was lucky enough to discover in Blackrock Cave during the 1988 expedition.

50 metres downstream from the sump a side passage was followed through boulders to Baby Sarawak Chamber measuring 100 metres in width, 77 metres above the river. Half way up the boulder slope a scree descent led back to the river but beyond was Cobble Valley with narrow draughting rift passages. Cobble Valley terminated up a steep slope at a choke of these slippery river cobbles. This was obviously a major inlet to the Clearwater River during times of heavy rain.

An ascending passage at the Sump Bypass between Clearwater V1 and V11 could be climbed past cave pearls and gour pools to emerge near the base of the pitch into the river from Armistice, a convenient bypass to Water Gate. It had been 8 years since any progress had been made in this river and to date this area has not been revisited.

This had been yet another of Matt's highly successful expeditions and he was to continue leading and organising Mulu expeditions into the far future. Tim Allen never stops and also organised and led many successful expeditions to Mulu over the years. Their discovery of Whiterock Cave with its connection to Blackrock extended the length of the Clearwater System to 238 kilometres. When we first began our exploration in 1988 it was 48 kilometres in length. Progress has been made.

8
The Japanese Expedition 1997

'Things turn out best for the people who make the best of the way things turn out.'

John Wooden

I could see from the start that this expedition was going to be a problem so this is a very short chapter as there is nothing much to say about it. The only answer for this expedition was to make the best of a difficult situation. With the past expeditions organised by the Indonesians and Koreans to Mulu with no results, maps or reports, I found it hard to understand why our Japanese friends had been given a permit to explore caves especially when it came to pass that they were all young students with zero experience of expeditions. A couple of the team members of around twenty boys and girls had some experience of adventure caving in Mulu as I had met them on a number of occasions and they seemed competent enough to get by. Considering that Japan purchases the majority of Sarawak's timber in raw logs I could see that it was probably difficult to refuse them permission. They were also engaged in making a film so did have a senior leader who spoke good English.

Their request was to concentrate on Gunung Buda but at this time the Buda Caves Project from the USA was engaged in organising yet another major expedition so I advised that they camp at Buda, cross the Medalam River and explore north Benarat for caves. We had a provisional camp there used for our Gunung Buda Caves Project work. Little work had been accomplished in this region of north Benarat by the British expeditions so I hoped this would keep them out of trouble. It was my job as usual to accompany them and gather all the data at the end of their expedition. This included surface and cave mapping information so I gave them instructions on how to use 'Compass' software for recording all cave measurements. This software also plotted a straight line and walls of the cave which could be printed out as a base line for the final hand-drawn map.

We settled into camp and I rigged a tyrolean rope across a shallow part of the river so they could gain access to north Benarat without drowning.

The valley floor was easy to navigate across as it's flat and free from swamps and a marked trail was established so they did not get lost.

Benarat north looking from Mojo Cave, Gunung Buda across the Medalam Gorge. Photo: Dave Bunnell

I also had two park guides from Mulu, Peter Tepun and Jonathan Kolay both experienced in caves and a few porters from the local community. Under their guidance we began searching for possible cave entrances but as usual with Mulu the karst is anything but easy to navigate across but we did find one of interest. This was a pitch around 40 metres above the valley floor sadly blocked after a descent on ropes of 20 metres.

They began a climb to find Gawai Cave and finding what they thought was Gawai mapped down for 1,775 metres finding footprints. This suggested that they had entered Blue Moonlight Bay Cave from this high level entrance. Due to the fact that no map was ever forthcoming from the Japanese we have no way of telling. Blue Moonlight Bay had been explored and mapped by the British Expedition in 1980 to 9.4 kilometres in length and extended in 2003 and 2005 bringing the total length to 9.7 kilometres. The Japanese cave survey 'Compass' plot can only be described as complicated. This area is confusing and recent exploration suggests that they may have been in a totally different cave. Gawai had been explored in 1984 by a British team with a length of 300 metres. It seems doubtful they would have missed a substantial lead which would be draughting. This large entrance is situated high up on the slopes of Benarat with difficult access. I suspect I spotted it from a helicopter with a tribe of Red Leaf Monkeys in residence a few years previously but was never certain if it was Gawai or another cave.

After a few weeks little had been found so attention was focused on a large isolated block of limestone to the west of the Terikan River. The local community knew of caves in this mountain and it had been completely overlooked by previous expeditions being separated from the main Benarat Mountain. We set up a camp of tents in a nice reasonably dry area near the Terikan River. The mountain did not have a name so I called it Batu Terikan. Later it was rechristened Batu Agung, Agung being King in Malay.

It was a pleasant walk from camp to the mountain with groups of Red Leaf Monkeys causing mayhem. We found plenty of caves but only one was given a name. The rest for the sake of convenience I called according to their position in the mountain.

The three Red River East Caves were developed at base level separated by flooded sections and mapped for 532, 234 and 129 metres respectively. To the west again a nice series of four caves which for want of a better name I christened Red River Caves West. A cave to the north was only short with a nice formation at the end which was filmed. I found it disconcerting when pools of water with cave aquatic species were trampled underfoot. The team missed a good going lead up to the side of the passage. Being alone I crawled upwards past some crystal lined pools and emerged in a large high level passage going in both directions and full of cave formations. Sadly the team did not return to explore and map and this remains unexplored.

Tobichi Cave was a typical classic Mulu Cave explored and mapped for 1,146 metres with nice formations. This cave was later resurveyed by Rambli during a Mulu Malaysia scientific expedition. Further south a

2,775 metre long cave was explored which I marked up as South East Cave No1. To the south west was another cave which was measured at 865 metres. In total 7.1 kilometres of caves in this mountain had been mapped and there is probably much more to be found.

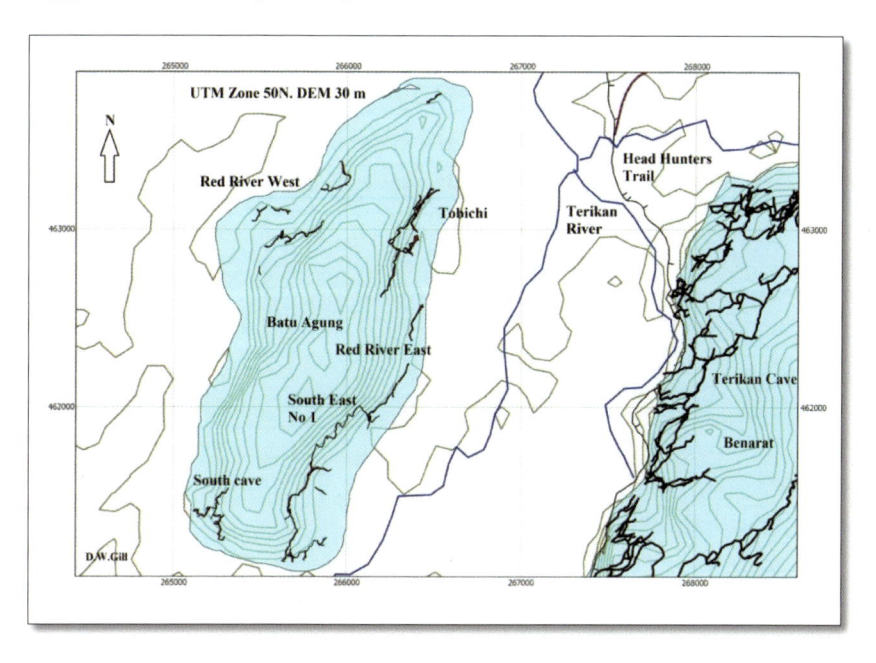

Caves of Batu Agung, Mulu National Park

So the expedition was not a complete failure as I did manage to obtain the cave survey data on 'Compass'. Sadly, after repeated requests, the report and cave maps were never received. This was a repeat of the previous Korean and Indonesian expeditions. It is gratifying to note without bias that all the British caving expeditions have produced and supplied well produced and informative reports and supplied photographs of superb quality. Interesting was the respect shown to my park guide Peter Tepun as they liked him so much they paid for his trip to Japan as a thank you present. Batu Agung really needs a revisit, the caves mapped, drawings and descriptions documented.

9
Gunung Buda, forming a new national park

'We don't meet people by accident. They are meant to cross our path for a reason'.

Unknown

As can be seen from the brief chapters, life in Sarawak was never dull as there was so much work to be done. Only a summary has been described as the work was often seven days per week so there was no time to become bored.

Gunung Buda. Photo: David Gill

It can also be gathered that expeditions in an attempt to discover caves was a lifelong love and especially in Mulu. Working in Mulu with a growing awareness, appreciation of the remarkable biodiversity, geomorphology of the caves and conservation could become obsessive and with a huge area of limestone to the north it was hard to resist the opportunity to attempt to constitute this area into a national park. Thus began the Buda Caves Project.

This magnificent block of limestone is a continuation of the Mulu National Park karst and was originally proposed as a part of Mulu. Unfortunately this was not to be as a certain government minister had a preoccupation with logging. Of course logging karst is against the rules as for one thing in the tropics it's rugged with thin soils and impossible to penetrate with a tractor. The other major factor is that destruction of the trees would kill the rooting complexes which in essence hold the poor thin soils together, the result being bare limestone or better described as desertification. The trouble was that the limestone is surrounded by productive forest. The area was a part of the Medalam Protected Forest and this means that it is state land under the jurisdiction of the Director of Forests. This basically prevents any of the local community trying to claim the land. The Medalam River is the dividing line between Mulu and Buda, the beautiful Medalam Gorge being the dominant feature.

Before a proposal to constitute the area can be compiled justification is necessary which means intensive study on its biodiversity and in the case of Buda its geomorphology including any caves. Only when the study is complete can a proposal be submitted.

In 1980 Buda was visited and a few caves mapped during the Mulu expedition by the late Dave Checkley, Colin Boothroyd, Nick Airey and Tony White. Dave commented in his report that he suspected a large cave system to exist due to the resurgences at Beachcomber and Turtle caves, the source of the Buda River.

To make a start we organised a reconnaissance from National Parks and Wildlife Department under the direction of Oswald Braken Tisen, my boss. Baei Johnny Hassan one of my park guides, knew the area well as his father collected bird nests from one of the caves.

We had a small Guard House within the Mulu National Park at the junction of the Mentawai River where it joins the Medalam so had somewhere to spend the night. Access into the area in 1992 was not easy as there were no roads so we travelled with great difficulty by longboat, upstream on the Medalam River to reach the Medalam Gorge at Gunung Buda. The majority of the time it was a case of cutting fallen trees blocking the river and pushing the boat. We landed on a pebble beach close to the mountain; this was used as a vehicle washing area by

Goodwood Logging Company. A logging track had been constructed along the gorge which gave us access to the mountain.

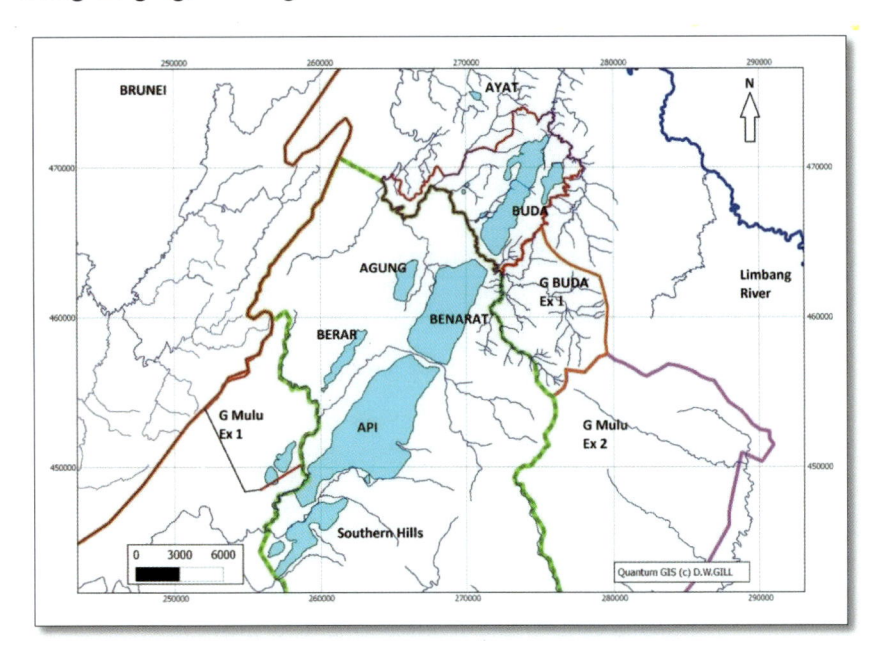

The karst area of Buda in relationship to Mulu

As time was limited we concentrated on the base level of the limestone and found Quill Cave to the east, only short but with some nice cave formations. Baei led us up the cliff using tree roots for hand and foot holds to old burial caves situated on a small ledge. We also just had time to check out Turtle Cave, a large cave with a river and deep water leading to a sump after 700 metres.

It was obvious that this magnificent limestone mountain deserved total protection either as an extension of the Gunung Mulu National Park or as a national park in its own right.

In order to establish a totally protected area in the tropics it's vital to understand the problems of enforcement likely to be faced. First and foremost, what is the objective that we are trying to achieve? Conservation of the biodiversity, which in a tropical karst area is high, conservation of its outstanding natural beauty as a karst area in the tropics is generally spectacular. Also of high value is the conservation of the geology and geomorphology of the karst and caves. Generally the caves in the tropics are outstanding in dimensions and are highly decorated with calcite formations.

Much has been written by academics for the International Union of Conservation of Nature (IUCN) as regards to karst and cave conservation but the majority of the papers are irrelevant to the real issues facing managers of these areas. An understanding of the science and a visit of a few days is hardly sufficient as the only way to gain an understanding of the conservation problems is to live and work there with the local community for many years. It can be seen that my work in Mulu as described in previous chapters taught me much on the culture of the local communities with the growing understanding that, on the whole, they were subsistence farmers with little or no income. It was normal to kill wild animals for food supplies thus poaching would continue even if the area was scheduled as protected. Fencing of the area would not work along with armed enforcement officers.

The Penan community are a typical example of the problem facing conservation. The forests are logged resulting in the loss of the wildlife. Logged forests will regenerate providing the animals are in sufficient numbers to disperse seeds but conversion to palm oil plantations results in complete irreversible destruction. Should the Penan be given large areas of primary forest as their own and regarded as an anthropological curiosity or an attempt made to integrate them into society? The decision made was integration which would take many generations to achieve.

The only solution to conservation of a tropical karst area is the concept of peoples' parks where the local community manage and protect the area with all income distributed to the local community via a cooperative. This system would provide a steady income implying that the long term conservation of the karst and caves was in their interest.

It was a case of beginning from scratch, compiling all known data on the Limbang District including the local populations, getting to meet them and the understanding of their culture, dreams and aspirations. What did they want? It was also vital to gather as much information as possible on Buda's biodiversity, geology, geomorphology and its caves.

The Limbang District of the 5th Division has an area of 3,976 square kilometres with a population of approximately 50,000 plus. This equates to 12.5 persons per square kilometre. Limbang District has only one Sub District of Nanga Medamit and Buda lies in this Sub District. Including the second district of Lawas the population of the 5th Division is around 94,000 with a total area amounting to 7,790 square kilometres.

Medalam River. Mulu on the right, Buda on the left. Photo: David Gill

It's interesting to note that a major part of the Gunung Mulu National Park lies within this 5th Division as the division boundary follows river catchment areas. From the Melinau Gorge at Camp 5 the river flows south to the Tutoh and Baram Rivers so is a part of the 4th Miri Division but just a few metres from the Melinau River are a few streams flowing north to the Medalam and Limbang Rivers, thus Benarat Mountain is within the Limbang Division. This is not a problem as the park is administered by the Forest Department under the specific Sarawak laws.

There are 14 Longhouses within 25 kilometres of Buda with groups from the Tabun, Iban and Murut communities. The Tabun are closely related to the Berawan from Mulu and were the original inhabitants with their residence at Kuala Medalam, Kuala meaning a junction of two rivers, in this case the Medalam where it joins the Limbang River. The Iban groups were originally from the west of Sarawak in an area close to the Indonesian border of Kalimantan. Due to the troubles in the 1950's with the Indonesian invasion during the British administration, the Iban were moved to safety on the Medalam River so are thus regarded as immigrants. There has over the last 70 years been much intermarriage so animosity between the different groups is slight. The Tabun therefore considered Buda to be in their territory and would lay claim. This over the years has caused much misunderstanding with regard to establishing a national park.

All I had to do now was reach agreements with the local communities and gather all the data required to justify a new national park, a mammoth task.

George Prest, a caver from the USA had met up with Paul Norman at Lechuguilla Cave. Paul had caved in Mulu with Alan Weight's expeditions so told George all about the delights. He recommended that if George wished to explore caves in Mulu he could join a British expedition. Later George Prest and his friend John Lane had read the Mulu reports of the large, beautiful and extensive Mulu caves and of course wanted to visit and explore. It was a dream which proved very hard to bring to fruition but the two were characters of the type that never gave up.

John contacted Tropical Adventure in 1993 and my predecessor Mike Meredith also gave them information with a recommendation that they have a tourist trip to the Mulu caves first. George wrote to the State Secretary and received a positive reply so they recruited a team of USA cave explorers but were unable to gain any sponsorship. At a meeting with Richard Hi of Tropical Adventure in the USA they worked out a basic price and logistics for the expedition.

They managed to obtain airline tickets to Borneo in 1994 to conduct a reconnaissance and met up with the Secretary of Tourism and Environment and the Sarawak Tourism Board. Attempts to obtain topographic maps failed as maps are not made available to the public in Sarawak. At a meeting with Braken he suggested that the USA expedition should concentrate on Gunung Buda as it was wide open awaiting detailed exploration as the caves of Mulu were systematically being explored by the British expeditions. George and John flew to Mulu where I met them to discuss the expedition objectives; this also gave them the opportunity to have a trip in the tourist caves and meet the team members of Alan Weight's expedition as they were there at the same time.

They flew to Limbang town with Paris, a local cave expert and my park guide Johnny. Again they failed to procure maps from the Limbang Forest Office but managed to obtain transport along the logging roads to the Limbang River crossing. This was a two hour drive but there was no bridge across the Limbang River so they hired a boat and scrounged a lift from a logging truck on the far side. One bumpy hour later they reached the Medalam Gorge where Johnny and Paris constructed a bivouac camp. George was feeling very sick and the following morning they looked at the small cave near the road at the eastern end of the gorge but George was unable to continue so Paris took him back to Miri where he eventually recovered. Three days later John returned from Buda with good news, he had found six other cave entrances all looking good.

Back in the States Richard asked them to give a presentation for the Malaysian Promotion Tourism Board in Los Angeles. They spent all night producing 5,000 brochures and paid for an Expo booth and it was not long before their efforts started to pay off with sponsorship deals including equipment. This had entailed a mammoth amount of work over a period of two years. On 26th December 1994 the team set off to Sarawak, the first USA expedition to this part of the world eventually leading to the justification I required for a new national park.

10

USA Gunung Buda Expedition 1995

'Most men lead lives of quiet desperation and go to the grave with the song still in them'.

Henry David Thoreau

For the sake of diplomacy it was decided to call this a joint expedition between the USA speleologists and our Mulu team as we had decided to do the same with the Mulu British expeditions. My job was to monitor, take part and gather all the recorded cave data. The permit for the expedition was issued by the Director of Forests Sarawak for a very strong and experienced team consisting of 16 from the USA and Alan and Becky Weight from the UK. On the Malaysian side was Simon Sandi, my boss and Louise Korsgaard from Denmark plus an assortment of Mulu guides. I decided on a rotary system for the guides as they could claim an extra travelling allowance, so as many as possible could participate. As they are all competent in caves and in the forest this was a great advantage to the expedition. I had some trepidation about working with a USA team I did not know but it turned out they were all nearly as mad as the British.

We set up a fine base camp in the gravel-cleared area where we had landed our boat during the reconnaissance and started exploring. The first cave to produce results was Cin Cin Cave named after the roly-poly insects. This was situated 100 metres above the gorge and a large entrance led to a 16 metre pitch. Western trending passages led to the free hanging 33 metre deep Croc Pot, so named as there were marks in the mud walls indicating a large animal with claws had been trying to climb out. The guides insisted it was the mark left by a crocodile. The cave was flooded beyond but the question was, how did a crocodile get in there and where was it now? A series of well decorated passages led to a lower entrance which proved to be a much easier access into the lower sections of the cave. The cave was mapped for 2.6 kilometres with a depth of 98 metres.

One of my guides Joseph Gau accompanied by Becky and Alan searched the cliffs at the eastern end of the gorge. They spotted leaves moving which indicated a possible cave entrance so Alan climbed up to find a one metre diameter tube guarded by a racer snake, beyond the crawl it looked like a large chamber could be entered.

*Shower Head and Rim Stone Pools in Cin Cin Cave.
Photos: Dave Bunnell*

As it was like a Tardis the cave received its name. The following day Simon and I joined the group which included Djuna Bewley and Dave Bunnell and thank goodness the snake had disappeared. The area was complex and the mapping began with another small entrance seen. I looked at a small passage which was draughting but it was a very low crawl leading upwards. I passed a squeeze to emerge in a massive passage heading off in both directions. Joined by Simon we headed off on a bearing of 60 degrees along the strike of the beds, the general direction of the majority of the Mulu caves heading into the mountain, the strike being at 90 degrees to the slope of the beds. At an area of large guano covered pinnacles we returned the way we had come to give the surveying team the good news. We were followed with much enthusiasm and Alan stole the best lead to the Pinnacles while Simon, Dave Bunnell and I mapped the passage in the opposite direction. We called it No Stone Unturned. This area was just as amazing with pinnacles everywhere but the massive passage terminated in a choke. We found another small entrance and unwisely decided to use it to exit the cave. One hour later we were still climbing down a sheer cliff using tree roots as hand and foot holds.

The following day we continued with the exploration accompanied by Louise and by climbing up through pinnacles followed the bedding steeply upwards. These ramps in Mulu often intersect massive high level passages following the strike but in this case all we found was a narrow passage heading in the correct direction. We mapped as we progressed to a vertical drop down requiring a rope and a bolting kit which we did not have. The narrow passage could be seen continuing on the far side of the pitch. Our mapping finished here and this area has not been visited since so can only be considered as a possible future lead into the unknown.

The Tardis Pinnacles. Photos: Dave Bunnell

Alan's team continued mapping north and they descended a pitch into more ramifications of passages that they christened 'Austin Series' as it took off from station number A40. We joined them with the mapping but Sarah Vieweg stepped on a pinnacle which gave way. Luckily her injuries were minor cuts and bruises from which she recovered from after a few days. This treacherous cave was not finished with us yet as we discovered another entrance from Austin Series this entailed a climb up loose boulders to gain the entrance and daylight. Simon went first but using a rock for aid it parted with his company and a huge boulder fall came crashing down as Louise and I dived into an alcove for cover. Again Simon's injuries were minor as by some fluke he managed to stay on the loose rock wall without falling down. Tardis has a few possibilities of being explored further by poking around in Austin Series into obscure openings. The cave was mapped for 2.64 kilometres with a vertical range of 128 metres.

In retrospect we came to realise how fortunate we had been as Simon and I were the only two people in the world to discover underground pinnacles of such immense proportions, never seen before in any cave system on the planet. The big question was-how did they form? It appears that small indentations are made into a rock surface by dripping water. Over many thousands of years these small indentations are filled with guano from swiftlets and bats increasing the acid content of the water. Alan and I had much fun drawing the map and trying to decide what we should call this new formation; Shit O Karst became the favourite.

I did not own a lap-top in 1995 but the USA team had one. This gave us the opportunity to use a cave survey programme called SMAPS where all the survey data legs could be entered. With this we could generate a straight line map which could be viewed on the lap-top screen, a great help when searching for likely leads within a cave system.

The Gunung Buda limestone was geologically complicated. The southern block had been partially truncated by a valley leading downwards from east to west. Thus the middle block was more isolated indicating that any cave systems in the southern block heading north would probably terminate in this valley. This central block also had a valley to the north again indicating the same problem. The third block had a magnificent steep-sided and narrow gorge to the north again dividing it as a separate block of limestone which would most probably terminate any cave systems heading north. The fourth northern block was huge with what appeared to be a large doline to the west side. An inlier also could be traced out on a map to the east side of this north block. This was where the 'Big Feature' was looked at by the late Dave Checkley's British team in 1980 as it's obvious from a topographical map. They were unable to descend this massive doline due to a lack of rope. It is possible that this inlier is connected with the north karst block via a steep-sided jungle-covered ridge but no one has ever been to look. There was plenty to go at as it was thought possible to set up a bivouac camp to the north.

There are two main risings flowing from the southern mountain, the source of the Buda River. The first is Turtle Cave, approximately 700 metres of deep water river to a flooded section. The source of this river was suspected as coming from the east-which we proved to be the case later. The second further to the north was Palm Beach Rising, the river feeding this rising was found in Beachcomber Cave. This was the area that the late Dave Checkley suspected as having an extensive cave system.

Turtle Cave River. Photo: Dave Bunnell

It was also interesting to note the cave mapping techniques of the USA cavers. They could spend 10 minutes at each survey station with immaculate detailed notes, unlike the British technique of straight line surveying. Of course it was important to leave a cairn with station number at all junctions, which enabled any further exploration to map to a known given point within the cave. Surface mapping was also vital to obtain the geographical position of entrances as GPS had not yet been invented.

Above Turtle Cave was a small entrance pointed out by Johnny's father who had a small hut outside the cave. This was an old phreatic cave system running above the river cave mapped for 3.81 kilometres. A few 10 metre diameter tunnels led eastwards.

The exploration of Snail Shell cave was something of an epic largely under the guidance of Herb Laeger. Besides being a highly competent cave explorer he was also a good climber with a British sense of humour. His team searched for a large cave entrance reported by the loggers high up the scree and jungle-covered slopes above the gorge. They found Pud Cave but it was only 85 metres in length. His Mulu guide compatriots searched along the cliff face and found a large cave entrance with little piglets at the entrance. Roland captured five and headed off to camp for dinner but later they were secretly released before they went to the barbeque. The USA cavers were conservationists.

The passage was large with a nice flat floor with many leads and another entrance. They called the cave Piglet. Herb returned a few days later with Don Coons and my Mulu guides, Jonathan Kulay, Roland Gau and Lawai Kumpeng. It was Friday the 13th which should have told them something. They mapped the large passage to a complete blockage but a huge ramp going as usual up dip was followed and 70 metres above the floor they arrived at a junction with three ways onward. The draught came from the left where to the right was a pit with no airflow. The third option was tackled by Lawai, a steep upward climb but on slippery ground he came crashing down, luckily catching hold of a passing stalagmite before he hit the floor. With cuts and bruises they retreated as this climb really required climbing equipment.

Don and Herb returned a few days later with Jonathan but the pit went nowhere, to the left was promising and led via a greasy climb to yet another entrance, the third. On the way back Don traversed the cliff to the east and found a huge entrance with lots of snail shells, thus the name, Snail Shell Cave.

A few days later Herb was back with Don, Lawai and Roland. They mapped a beautiful passage with massive formations and calcite stalagmites with pools of water and passages heading towards Piglet Cave. Finding Balcony Entrance gave them easier access into these passages and a connection was eventually made by Herb via a small hole in the floor of Piglet Cave bringing the total mapped to over 3 kilometres with multiple entrances.

The Bath Tub, Snail Shell Cave. Photo: Dave Bunnell

There were still plenty of potential leads and the draughting ramps proved to be the most spectacular which entailed numerous climbing trips. They eventually reached an amazing height of 470 metres above the entrance with a draughting lead being too narrow for progress. No doubt this ramp will pop out in a doline somewhere in the mountain. So Snail Shell became the cave with the highest vertical range in Borneo. The final mapped length came to 5.83 kilometres with a potential lead that could connect to Upper Turtle Cave.

Beyond Palm Beach Rising and Beachcomber entrance Dave Bunnell and Djuna were taken to a doline by Baie Hassan's father where a 20 metre wide and 5 metre high cave loomed off into the darkness. They explored to a canyon and descended a 14 metre pitch teaching Baie's father how to abseil. The passage beyond was full of pinnacles similar to those we had found in Tardis. The next day Dave with Jed Mosenfelder and Peter returned mapping like mad finding the Back Door entrance. There were so many leads and other entrances that it was hard to choose. Many kilometres of passages were mapped in this complex maze of a cave but one important event happened when Kyle Fedderly, Kevin Downey, Bradley Hacker and Jed Mosenfelder descended to a clean-washed river. Sadly it was sumped upstream and also downstream after a few hundred metres but the stream was running green. That day non toxic dye called fluorescein had been introduced to a cave on the east side of the mountain. This cave which they had christened Monkey's in the Mist had originally been noted by the British expedition in 1980. It takes a sizeable stream and is vertical for 140 metres to a sump.

The giant pendants of Biocyclone Cave. Photo: Dave Bunnell

Due to its narrow rift passage it had proved very difficult to explore with water everywhere, a little like caving in a flushing toilet. Yet another major breakthrough was Don Coon's descent into Pie in the Sky, the final point in Beachcomber Cave explored by the British.

The cave was mapped to 11.8 kilometres in length with passage heading south towards Upper Turtle Cave. It looked possible that the system was all one including Snail Shell with the potential of a 30 kilometre long cave system.

A bivouac camp was established to the north so this massive block of limestone could be explored. An important find was Pepper Leaf Cave, being only mapped for 180 metres it had a very large mature passage with many unmapped small side leads.

The Biocyclone Cave Pinnacles, the largest ever found.
Photo: Dave Bunnell

The isolated block of limestone to the east was investigated. The Sarawak Hotel was a huge overhang adorned with stalagmites and a reasonably dry area to bivouac. The Senap River was close by and the morning call of Gibbons was a nice alarm clock. Just beyond to the south was Biocyclone Cave. This was a strange old phreatic draughting cave with the largest wall pendants ever seen and pinnacles of huge proportions. The cave was mapped for 1.45 kilometres with many going leads remaining.

The Big Feature identified on the topographic maps was investigated by the British expedition but a lack of rope prevented a descent of the doline into what appeared to be a passage approaching the size of Deer Cave in Mulu. The cave was not easy to get to, involving ascent of a steep jungle covered ridge. They called it Spirit of the Jungle as a sign of great respect. The tunnel sadly terminated in a sump after 260 metres. Where the stream emerges is presently unknown but it is suspected that there are risings on the west side of the northern block.

In the early days of working in Mulu I found it necessary to appreciate the cultural differences as, one evening at Camp 5, with the guides we went for our bath in the river. Not thinking I stripped off and dived in but all the guides left their knickers on. After some stares from the guides it became obvious that this was not the right thing to do. The USA team of course did the same including the ladies much to the fascination of the guides. When a local lady has her bath in the river she is always dressed in her Sarung.

I came to realise more and more during expeditions to Mulu and Buda the expertise of my Mulu Berawan guides. They never seemed to get lost in the jungle, having been born and bred into this environment. They moved freely, breaking the occasional twig or marking the bark of a tree and often would glance at where the sun broke through the jungle canopy for a sense of direction. I tried to do the same but often failed. They were natural cave explorers and brilliant climbers. They would just follow their footstep marks to find the way out of these complex caves. I often glanced back to visualise the passage from the opposite direction and would leave a marker, usually a cairn, at junctions.

Many smaller caves were explored by this highly successful expedition with a total mapped of 29.34 kilometres, a brilliant achievement. There was no doubt the USA team would be back as there was much more to do.

The Gunung Buda USA team and the author, 1995. Photo: unknown

11

Gunung Buda USA Expedition 1996

'We need to be the change we wish to see in the world.'

Mahatma Gandhi

During the 1995 USA expedition John Lane and Todd Burks had the opportunity to meet with the Minister of Tourism, Datuk Amar James Wong. Datuk James was also one of the YB's for the Limbang District, the title YB being the UK equivalent of a Member of Parliament. Datuk James was keen to see Gunung Buda being constituted as a National Park and opened to tourists. This he believed would be of great benefit to his constituents. He kindly agreed to sponsor the expedition so John recruited a team of 20 from the USA of cavers and research workers. Also three from an adventure film company and five from the National Geographic Society to cover the expedition for the magazine. With five of us from the Forest Department and four from the Sarawak support company we had a total of 37 persons, base camp was a little crowded.
It was christened once more as a joint Malaysian-Subterranean explorer's' expedition.

Having some research workers was a great advantage as I could gather more biodiversity information in order to justify national park status. John was to conduct hydrographical studies and Adam Bodine looking at the archaeology in the caves. Steven and Stella Heydon spent their time collecting and identifying numerous species of butterflies while Ralph Cutter went fishing and snake hunting. He had one wonderful encounter with a huge Reticulated Python which he wrestled in the Senap River. Greg Milano was our main snake man with Nathan Schiff and Alan Mudge into entomology.

We were now using the cave survey programme 'Compass' so all our previous cave survey data could be entered into this programme. It's from the USA so all measurements were in feet but the default could be changed to metric figures. The main problem with all mapping in complex areas with many different caves was linking them all together so that a map could be created of the complete caves of the mountain.

This could only be accomplished in 1996 by surface mapping from a known point as accurate GPS measurements were impossible to obtain in 1996. The problem encountered was that, in order to link the caves, each survey station required a unique number otherwise two station names with the same number confused the programme and the straight line plot looked like a jumbled mess. The advantage with 'Compass' was that walls of the cave could be included so that the final free-hand drawing of the map became much easier. Volume could also be calculated and a special file created of fixed points so the map could be generated to fit on existing contoured maps.

Karst of Gunung Buda looking south towards Mulu. Photo: David Gill

As for the exploration of the caves we had Chris Andrews whose' first exploration was a cave up in the difficult-to-traverse karst block of the southern section of mountain. He explored in a large passage for 180 metres to another entrance with disappointingly no other leads onwards into the mountain.

Pepper Leaf was the large cave explored in 1995 but had many leads so a sub camp was set up close to the Senap River and Mark Deebel,

Greg Milano and Steve Reich began mapping the side leads. The main passage is 20 metres high and 15 metres wide but terminates in a mud fill. They mapped side passages with many loops arriving back at the same place. Later with two of my park guides they mapped to another two entrances bringing the total length to 1.1 kilometres.

This sub-camp was reached by following a logging trail to the north skirting the Buda limestone and then crossing over the Senap River via a dodgy collapsed bridge composed of one solitary slippery log. The Sarawak Hotel was merely a massive rock shelter adorned with stalagmites situated on the east side of the isolated block of limestone to the east. The problem was getting there, as a ridge needed to be climbed upwards again along an abandoned logging trail. Just before it hits the limestone a very steep descent is made back to the Senap River. Following the Senap River bank upstream from the sub camp was attempted but flood water made a good attempt to wash the explorers away.

From the Sarawak Hotel sub-camp Biocyclone was explored by Chris Andrews, Steve Reich, Greg Milano and James Colborn where they mapped a further one kilometre in the Coliseum Room but, although I obtained the survey data, no map was ever drawn up. With great difficulty they also managed to get some filming done with the film team. They reached the Sarawak Hotel by borrowing the helicopter and were deposited on a logging road to the east of the Senap River. With the film team carrying large heavy boxes of film equipment they had fun finding a way across the Senap and hiking through the jungle to the Sarawak Hotel. The group also explored Conair Cave at 120 metres in length and Coliseum Cave at 400 metres in length but, again, no map or survey data became available. The Big Feature, Spirit in the Sky explored during the last 1995 expedition, was looked at but no bypass could be found to the terminal sump.

Shelter in the Sky was aptly named as it's an obvious large cave entrance in the gorge cliff face which can be seen from the logging road. Chris Andrews and Todd Burks completed the 70 metre climb with four separate pitches. The entrance was spectacular with magnificent views of the north Benarat Mountain across the Medalam Gorge but, sadly, the cave only extended for 180 metres; but was filled with cave formations and phyto-karst. These are small pinnacles pointing towards the light formed by photosynthesis. Basically this is cyanobacteria growth which produces carbon dioxide and, combined with water, produces carbonic acid which dissolves the limestone. A huge swarm of bees seemed to be attracted to Todd who received numerous bites and was not feeling up to his usual self for a few days.

I had to return to Miri to hire a helicopter for Stephen Alvarez of National Geographic and flew back with the pilot. We obtained some nice photographs of Buda and Benarat. Even Datuk James popped over one day in his hired helicopter just to say hello.

The problem of guide rotation was an issue as the three park guides of Jeffery Simon, Thambi Epoi and Ipoi Lawing were so good at finding and exploring caves that John understandably wanted to keep them. This did not give my other guides a chance at joining the expedition, so missing out on their travelling expenses.

Pirate Cave was a strange place as it is situated in a small isolated block of limestone in the swamp forest. It's a maze of passages both vertical and horizontal. Another cave close by in the same block was called Birthday Cave but did not connect to Pirate Cave being largely developed at base level by a small stream which has undercut the limestone block. We could not understand how this small limestone block was still standing as the caves were mapped for just over 2 kilometres, it was basically hollow. In later years, on drawing the boundary of the planned national park, we extended the suggested boundary to include this area.

Adam Bodine's work on the archaeology was useful but although a few caves had been utilised in the distant past as burial chambers it was thought that any information on their geographical position should be kept as confidential. There were no archaeological treasures of any significance.

The North Gorge is a beautiful spot, a canyon with steep 500 metre cliffs no more than 50 metres wide. A small stream runs through the gorge which truncates the limestone, an ideal tourist attraction. To make matters more interesting it has three caves. Little Man Cave was disappointingly only 191 metres in length while Daylight Cave was even shorter at 34 metres but the most interesting find was Dawai Cave mapped for 300 metres in length. This cave contained a profusion of spectacular cave formations including helictites, crystal-lined pools, soda straws and shower stalactites. Two very small entrances were found approximately 25 metres above their bivouac camp below a balcony. The main way on was to the east which was not fully explored and is worth further work. This cave would make a nice tourist attraction. Sadly I never managed to get to Dawai Cave to check it out.

Pepper Leaf and Snail Shell Caves. Photos: Peter Bosted and Dave Bunnell

The big find was in Snail Shell where Chris and his small team looked at the lowest point of the cave which according to the 1995 survey was flooded. Chris waded through a large 6 metre by 4 metre lake up to his chest with mud walls and mapped onwards for one kilometre along a huge river passage terminating only 300 metres from the Turtle Cave sump. The first section obviously became flooded to the roof during heavy rain. They called the find A River Runs through It; the drainage being along the Medalam Gorge, east to west. A dye test was conducted later and the flooded connection confirmed as the Turtle Cave River turned green.

They were not finished with Snail Shell so studying the 1995 survey found other areas that remained a mystery. A 10 metre climb was made by James Colborn and the Mighty Skipper Borehole mapped along a 15 metre by 20 metre passage for a distance of 500 metres. The final climb involved three trips bolting up a mud-covered calcite slope for 70 metres completely closed down at the top. The surprising find was the three 'Blister Balls', 2 metre oval concretions like black shelled eggs composed of soft white calcite beneath the shell. They could be ancient decomposed cave pearls of a remarkable size. As far as is known this special formation has never been seen in any cave previously. Snail Shell had been extended for 2.8 kilometres bringing the total length to 7.36 kilometres.

North Gorge. Photo: Andy Eavis

The expedition had been highly successful with over 8.1 kilometres of caves explored and I now had a great deal of further information on the biodiversity of this area in order to formulate justification for a new national park.

I was asked to lead a small group from the National Parks and Wildlife

Department, (which included the State Secretary's son), to Gunung Buda to look at the caves and the swiftlet populations. I was not well at the time with a lung infection and finished up in hospital with pneumonia.

Spirit of the Jungle Cave, Gunung Buda. Photo: Dave Bunnell

12

The Gunung Buda Caves Project 1997

'Those who dare to fail miserably can achieve greatly.'

John F. Kennedy

The following year Joel Despain and George Prest came back to their beloved mountain, this time with a total team of 20 for a planned 36 days. I had 8 park guides involved on rotation plus Simon Sandi and me. The name became the established Gunung Buda Project, again a joint undertaking with the Forest Department. We could now cross the Limbang River without boats using the logging company suspended winch with its small cage for us to clamber in. Our base camp was established in the usual place near the Medalam River offering a nice spot for a swim and to get clean after a long day underground.

The maze of Green Cathedral Cave was the main focus of attention as in 1995 the cave had been connected to Beachcomber Cave bringing the total length to 11.8 kilometres It was also close to Upper Turtle Cave where a hoped-for connection was possible.

One lead in the northern section of cave explored and mapped by Joel, Merrilee Proffitt and Bradley Hacker led to a 3 metre diameter stream passage which terminated in breakdown and, from the survey, appeared to be beneath the valley which divides the southern block from the central area. Due to the proximity of this valley it seemed doubtful that Green Cathedral could be extended further to the north unless a way could be found beyond this chaotic breakdown beneath the valley.

The trip to Snake Heaven-so named as it's guarded by a Racer Snake-was an exceptional piece of cave exploration. Joel Despain took Greg Stock, Stan Allison and Kyle Fedderly to the lead which was an enormous draughting hole in the roof with overhanging boulder-strewn walls. It could only be described as an intimidating looking climb. The first 15 metres of the walls were overhanging with mud covered ledges but a good ledge could be seen about 25 metres above the boulder-floored chamber. On the way out Joel was surprised to see the same

Racer Snake there that he had spotted two years previously.

Joined by Mark Fritzke but without Joel they returned a few days later with some climbing gear which was limited to a short rope and a few expansion bolts. They spotted a small natural bridge approximately 12 metres above and for the next four hours with tremendous persistence they attempted to throw a thin string line tied to a rock across the bridge. Eventually it worked and a pull through of a rope allowed Kyle to ascend to a traverse requiring a few bolts for a fixed line. Stan then took the lead and, by using a lassoed projection, reached a muddy slope. Mark then proceeded up the climb by kicking foot holds in the mud and finally reached a horizontal passage which was only short in length. The draught was still blowing from above with the way on across the shaft and up another climb. They headed out at 10 pm but returned the following day. Stan traversed across the drop using a few bolts for a fixed line then Greg proceeded yet further upwards and rigged a static rope to the floor of the chamber so all could follow without risking the traverses. At the top Snake Charmer Hall was huge at 60 metres high by 30 metres wide with two waterfalls cascading down from way above. Beyond was an array of cave decorations with white calcite walls but no direct way through could be found when they returned a few days later to complete the mapping. The wind seemed to come from way up which required a substantial long climb and much equipment. It was decided not to leave the ropes in situ so the climbs were sensibly de-rigged. Any further exploration beyond Snake Charmer Hall is going to be a serious proposition.

One of the top priorities for this expedition was to establish a connection from Green Cathedral to Upper Turtle Cave. I had procured a couple of bottles of California Red wine to be kept as a celebration. Greg Stock, Herb Laeger, Mark Fritzke and Ron Simmons knew of a good-looking climbing lead in Beachcomber Cave. Herb had little problem free climbing a couple of steep ascents, fixing a line for the others to follow, arriving at a chamber with three ways on. Two of the ways on led to pitches so they followed the horizontal passage which led to an area of Moon Milk formations and tree roots with many entrances to day light. They named it to save confusion as the Sink City Entrance Complex. It looked promising as the passage was heading south west towards Upper Turtle Cave. They used this entrance as a much easier route back down the karst rather than the long way underground.

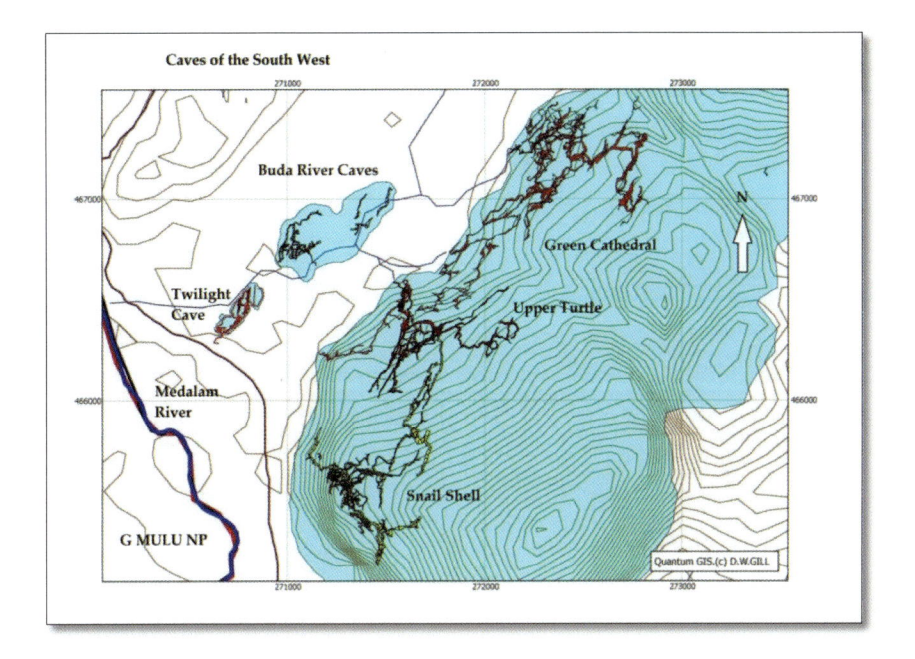

South west cave systems, Gunung Buda karst

The following day they were back armed with more men. Stan Allison, Joel Despain, George Prest and Mark Rosbrook joined in the fun. Mark descended a 30 metre deep rubble-strewn pitch but it was a blind lead; however Greg found a Moon Milk-covered slide down a tube which looked promising with a chamber containing a profusion of broom stick stalagmites. The second team had descended the two pitches found the previous day but again both were blind leads. A few days later Greg, Mark and Herb were joined by Kyle Fedderly and the entrance series produced a surprise as, hanging from the tree roots from the ceiling; they found frog eggs with a fully developed frog inside. This tree species by-passes the tadpole stage and hatches to a complete tiny frog and then climbs the tree root to the cave wall. It's a way of avoiding predators. They found a by-pass to the Moon Milk Slide and from the chamber with the broom stick stalagmite they mapped onwards to a un-Buda like crawl but as it was draughting they pushed upwards in a narrow muddy passage. Herb popped out into a large passage at a junction. They headed south west and entered a 10 metre by 15 metre tunnel. Mark recognised it as one of the Upper Turtle Cave tunnels. They mapped to station number HA 19 and celebrated the long-sought connection between the two caves. Back at base camp they kept it a secret until they entered the station data into the laptop when there were some surprising looks. The California Red did not last long. The combined system was now a very respectable 21 kilometres in length.

Cave Racer, Elaphe taeniurus says hello. Photo: Dave Bunnell

Snail Shell Cave still had leads that required a good look at and Merrilee Proffitt, Ron Simmons, Don Coons, Herb, Greg and Stan attacked a climb finding a large passage eventually leading to another entrance to this complex cave. They used this to climb back down the mountain, as usual with great difficulty trying to negotiate the highly dangerous karst terrain.

The most horrific story came from George Prest, accompanied by Don, Greg and Merrille; they tackled a climbing lead which led to a muddy draughting constriction. Merrille being small managed to negotiate the squeeze and reported large passage beyond. Many hours were spent enlarging the squeeze so that the larger members could get through. George was the last but became totally stuck thus blocking the escape route for his three companions. After much struggle and pain and after a long period of time George thankfully emerged on the other side with his caving clothes torn to pieces. They mapped onwards and found a previous survey station and a blank wall where there was supposed to be a rope hanging down. This escape route would be the obvious route for George to take as getting back through the squeeze was considered as impossible. Merrilee and Greg had no trouble with the squeeze so returned with a rope to tackle the blank wall. One hour later water now started to pour down the shaft as it was obviously raining heavily on the surface. Don and George began to get rather worried but Greg at last appeared at the top and rigged the pitch so escape was now possible. Back at camp the state of George caused much amusement but he considered himself as a very lucky man.

Snail Shell Cave. Photo: Dave Bunnell

Further mapping added 3.5 kilometres to the total length of Snail Shell bringing it to 10.8 kilometres-but a sought-after connection to Green Cathedral and Upper Turtle Cave remained an anomaly. The usual talk was of 'bore hole', a term used by the USA cavers but not by the British.

We had a Malaysian film crew with us documenting Gunung Mulu National Park and wanted to include Buda as a part of the production. They were busy obtaining excellent footage of Snail Shell Cave when the camera man along with his assistant slipped and fell down a muddy hole in the floor. They landed 10 metres down into water and mud and the concern was, do we have two badly injured or dead men? A shout indicated they were relatively uninjured and more importantly still alive. The rather expensive film camera was in a back pack and went down with them and they retrieved it from the stream. My park guide Peter Tepun took charge and raced off the long way back to base camp for a rope. He returned an hour or so later with two more of my park guides and Robert Gani descended to help them out of their predicament. Bruises and cuts and a destroyed camera proved to be the ultimate result.

A major find was Loris Cave high up above Upper Turtle Cave. Some desperate climbing using tree roots for aid found Djuna and Damian Ivereigh at an unvisited entrance. They were joined a few days later with Joel finding the way on beyond the confusion of the entrance series. The cave was a strange one running for 1.3 kilometres over the top of Upper Turtle Cave but no connection could be found between the two caves. It had a number of entrances and one led to a massive depression in the karst. In this doline Djuna found a cave entrance with a draught but time did not permit a return. They called the doline the 'Lost World.' Loris was obviously a part of the old river system long ago abandoned due to the uplift rate of the mountain. In Mulu the uplift rate had been calculated using paleomagnetic dating techniques. This involves checking the direction of magnetic particles in sediment as we know that the earth has reversed its polarity a number of times over millions of years and the dates of the reversals are known. Andy Farrant's work in Mulu established a rate of 19 cm per 1,000 years for the uplift rate so depending on the altitude of the cave passage the age could be calculated. One sample of sediment from Cin Cin Cave at 152 metres gave a reversal indicating an age of 910,000 years.

I have many happy memories of exploring caves in Gunung Buda; one was a visit with the team to the late Minister Datuk Amar James Wong's house in Limbang for Chinese New Year. The Dragon dance was as usual impressive and the Minister gave me instruction to leave Mulu as I had done a good job there.

'Concentrate on Buda for me and obtain national park status as I want a national park in the Limbang District which will bring benefits to my constituents,' he said.

I did not argue as I was already fully committed.

My wife Betty came to stay at the camp for a week or so which was wonderful to see her around. I took her up to one of my favourite spots, a huge sandy dry floored overhang on the west side of the mountain with stalactites in the roof. I called it the 'Buda Lounge' as it even had a cave. The cave was only a small descending passage with little draught so it was never explored to a conclusion. We bivouacked there for a few nights just the two of us sleeping under a mosquito net and cooking over an open fire. The night time jungle noises were hardly disturbing with just wild boar and deer rummaging in the forest and not a snake to be seen.

Betty joined the national park guides and me in mapping Fruit Bat Cave. This was situated in another small block of limestone next door to Pirate Cave (mapped on the previous expedition). It involved cutting a difficult trail for 500 metres through swampy jungle but the cave was worth it with lower stream levels and high level passages. With nice formations it could be used as a tourist trip as it contained huge pinnacles similar to those we found in Tardis Cave in 1995. We mapped it for 905 metres but the map was never drawn up as the guides were supposed to do it. Again it appeared that this block of limestone was virtually hollow.

One of my favourite caves we explored and mapped was Twilight Cave formed in a small isolated block to the west of the southern block. It was hard to imagine there would be a substantial cave there but indeed there was. Andrea and Mike Futrell can only be described as mapping fanatics; Mike's cave drawings are that accurate in scale and direction they are good enough to publish once you wash the mud off the notes. The cave is a maze of interconnecting passages with streams and for considerable distances only has one wall with the other being in daylight, thus the name. It reminded me of a nice typical stream cave in the Yorkshire Dales but hotter with tropical gardens. This is the kind of cave you could happily take tourists through along with their children. We mapped 1.1 kilometres with two walls and 331 metres with one wall, total length an amazing 1.4 kilometres. The Buda River had certainly meandered around.

Mojo Cave or the Magic Man was discovered above Cin Cin Cave but only 229 metres in length. It was well decorated and terminated in another entrance higher up with an outstanding view of Benarat and the Medalam Gorge. Again I thought it could be used as an adventure caving trip for tourists. A connection to Cin Cin could not be found which at first looked likely.

The discovery and exploration of Deliverance Cave was cheating as the expedition was intended to concentrate on Buda. The north side of Benarat Mountain was inside the Gunung Mulu National Park but, as none of the Mulu expeditions had looked at this area, it seemed like a good idea. Herb asked me permission so of course I said yes but don't tell anyone. They went for it and told everyone. With Ron, Djuna, Mark Rosbrook and Vivian Loftin they crossed the Medalam River as the river was in low flow and climbed 200 metres up through the steep primary forest to the base of a cliff. They split into two groups and Herb went to the right only to find it became too steep for comfort. They returned to find the second group's kit near an entrance that went nowhere. A second small entrance beyond was blowing air so they squeezed through to find the rest of the team in a huge 20 metre high and 60 metres wide highly decorated passage with shower heads pouring in water from the roof. They scooped beyond directly south and into the heart of the mountain arriving at a huge room with a 40 metre vertical drop down and a waterfall pouring from the roof. The spray obscured their view of the far wall. Without a rope they returned to camp with big smiles.

Of course all team members wanted a taste of this discovery so, to make it fair, they devised a series of mapping expeditions so everyone would enjoy the opportunity. The following day which was Valentine's Day I joined the group as I was not missing out on this one.

Along with park guide Philip Lawing we mapped the massive entrance passage full of beautiful popcorn formations and rigged the drop which turned out at 37 metres. The room at the base of the pitch measured 65 metres wide, 45 metres high and 65 metres in length and received the name 'Dream Arena'. Mark, Vivian, Herb and I started mapping the passage on the right while Philip, Ron and Djuna went straight on to the south. After 400 metres we finished up in a dead end but magnificently decorated with calcite formations. One find was a snake skeleton covered in pure white calcite crystals, thus the name 'Snake Bone Alley'. Mark had a nasty fall on loose rock and picked himself up covered in blood, cuts and bruises. Was this a message of things to come, 'The Night from Hell?'

Showerhead Ballroom, the entrance series of Deliverance Cave.
Photo: Dave Bunnell

We met up back in the large room where Ron's team had mapped onwards in a huge strike-oriented passage 30 metres wide and 50 metres high noting a number of side leads. We headed out with our injured Mark who just kept going but on emerging to a dark and dismal night it was pouring with rain in the rainforest

With difficulty we navigated back to the river but it was in severe flood, there was no way we could cross back to base camp. The powerful current of the Medalam River was sweeping huge trees downstream, a swirling mass of brown raging water with massive standing waves. We tried a few times to ford a crossing but it was suicide. Eventually we gave up trying and trudged on through the dense undergrowth at the side of the riverbank. Philip led the way as we progressed slowly down stream towards base camp, which I judged from a previous trip to be no more than two hours away. As usual the Gunung Mulu National Park and the Sarawak jungle threw up a few surprises; large poisonous snakes, stinging plants, biting bugs, the usual sort of thing. There seemed to be large numbers of fallen trees that I could not remember being there before. We were tired, wet, cold, and hungry and our lights were fading fast. We had managed to explore over 1.5 kilometres of massive cave passages so we were all running on adrenalin, on something of a high which I suppose helped to keep us going. Also the thought of some food, dry cloths, a sleeping mat and the tales we could tell drove us on. By 3 a.m. we had all had enough, I judged that base camp was still a long way off and we still had to cross the raging river somehow. Thoroughly exhausted we collapsed beneath a small overhang, which offered a little respite from the rain and the leeches. I tried to sleep but the sand flies were out in force. My feet were a mess; Mulu Foot-that dreadful fungal infection of the feet which had the nasty habit of creeping up on you, meant that every step was akin to walking on hot needles. The USA contingent decided that the horrors of the Gunung Mulu jungle at night being far preferable than the bite of a thousand sand flies, decided to push on. Philip and I dozed off into fitful dreams. What was I doing here? Did I really get paid for doing this? What a strange occupation to have. Whatever made me throw in the towel in England, throw in my lot to the Borneo jungle and give my commitment to Sarawak, the Land of the Hornbill?

The next few hours for the five USA team members became an epic, one not to forget. Covered in leech bites Herb and Ron attempted a crossing but lost their footing and were swept away. They were still carrying their sacks which helped to weigh them down. Herb grabbed a passing small tree trunk but was threatened by being swept beneath ensuring certain death. He hung on with one arm and managed to get his legs around the tree and saw a light shining underwater and moving fast. It was Ron drowning so he grabbed his caving suit and held on. Ron did the same loving the tree passionately. Herb released his sack and managed in a desperate struggle with one hand to release Ron's. Both sacks disappeared downstream. With waning strength they hung on and amazingly Djuna and Mark found them and-using a rattan hanging from a tree as a line-threw it to them. Herb grabbed it and was washed towards the bank which was only 5 metres away. Ron did the same; he had lost one of his boots and 1,500 US dollars worth of camera kit. They bravely trudged on arriving opposite the camp at 7 am. The force of the river had reduced sufficiently for them to cross with help and encouragement from the other team members. Mark christened the cave Deliverance.

Philip and I had a very uncomfortable night but come daylight around 6.30 am we made an attempt at a crossing. Thankfully the river had reduced in ferocity. We just managed to get within a few metres of the opposite bank and Philip's huge brute strength made it. Throwing me a length of rattan vine I was washed ashore to safety. For future trips we rigged a tyrolean across the river.

Numerous mapping trips continued in Deliverance during the entire course of the expedition. The main south west route led into the huge tunnel of The A Passage and Gnome Hall and un-expectedly terminating in a large chamber they called Fossil Guano Dunes. This was a strange place with windblown stalagmites indicating that at one time this tunnel continued. A high passage was noted up in the wall. From the 'Dream Arena' a well decorated series of passages looped back to the main route but a side lead again had a huge high passage, a possible climbing lead for the future. Herb and Stan attacked a passage leading north to a traverse where they used bolts drilled into the wall. They called it 'Revelation Traverse', maybe something to do with Deliverance. This led them into a massive room, 110 metres by 55 metres and 40 metres high but it was a blank wall all the way around. One trip included sleeping underground for one night and another involved Merrilee and Greg as on returning they floated down the Medalam River on rubber tubes and back to base camp. Surprisingly they found Ron's bag. The total mapped came to 3.57 kilometres.

This expedition was not over yet, there was still much to do. Biocyclone received a visit extending the cave length to 1.45 kilometres with still plenty of open passage left to map and explore.

Vivian and Joel using the North Camp explored the country to the south following a small stream. Heading off into the karst they found Lizard Cave and not far away another draughting hole they called Langur Cave but the big find was Hornbills' Secret Garden Cave. This was situated at the end of a 500 metre long doline but required a rope descent to enter. They returned with Mark, Dave Bunnell and Bill Frantz and descended a 10 metre drop into a passage going in two directions. Over a few days this find was explored down to a sump and on the last day a huge 40 metre wide tunnel with a daylight entrance 60 metres above. The passage beyond was 10 metres wide and the exploration was halted at another pitch awaiting the next expedition. 1.9 kilometres in total was mapped.

Langur and Lizard Caves are separated by 30 metres with two entrances each. Both caves were mapped to around 330 metres in length. As with Pepper Leaf these caves are full of river sediment and river cobbles.

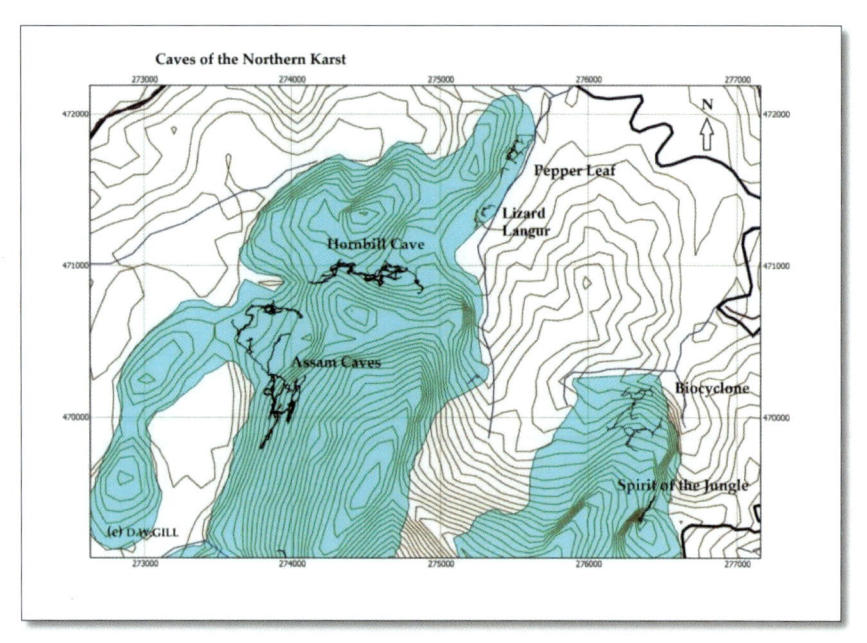

The caves to the north, Gunung Buda

The 1995 Sea Breeze Cave was once more visited in an attempt to explore all the way through the mountain to the west and Green Cathedral. They finished up losing much skin and clothing without a great deal of progress.

On the first day of the expedition Merrilee, Don, Heb and I set off on a long trek towards the North Gorge under the leadership of a local bird nesting man who knew the whereabouts of a cave high up in the cliff. As usual this was a case of climbing using tree roots and twigs for aid with lots of crumbling rock. After a hairy traverse using rattan for aid we reached a nice looking entrance of 12 metres by 12 metres. We explored onwards through some chambers to a black space where we could not see the walls. Don wanted to return tomorrow but I shot off down the rubble slope into the darkness. The chamber I called 'Limbang Ballroom' although it's not the place you would want to hold a dance. The mapping produced a survey of 80 metres by 110 metres, the largest chamber so found in Buda but all walls were blank. We noted 7 passages high up in the east wall, climbs for the future explorers as they should lead up the beds of limestone to high level passages trending along the strike. Later Don completed some side climbing leads and discovered a high up sky light. The cave was mapped to 611 metres in length.

It was decided to begin exploration of the west side of the mountain to the north by establishing a West Camp at Disappointment Cave. This cave had been partially explored by the late Dave Checkley and party during the early Mulu expeditions. The team split into two parties and began mapping Disappointment but Merrilee and Brad Hacker became frustrated in mapping tree root-festooned passage basically following the cliff line. They decided to find a cave of their own and, thrashing through the karst undergrowth, found a small blowing hole. This cave kept the team busy for some time and they gave it the name Babylon. Attempts to connect Disappointment to Babylon failed but the system was mapped for a very respectable 3.6 kilometres. Sadly, along with Hornbill, I did not get a chance to help with the mapping of these two caves.

Once more this had been an amazingly successful expedition with a wonderful dedicated group. 25 kilometres had been mapped bringing the total of explored caves to 54.3 kilometres.

It was not all good times at Buda as there were one or two disconcerting events. The first was helicopter logging. This involved logging on steep ground as driving a road and skid trails for the logs to be extracted are a time consuming exercise and expensive for short term gains. The steep mountain bordering the catchment area of the Medalam River was still under logging licence. I took a look at this mammoth helicopter machine in operation on the mountain. It was owned and crewed by Russians and appeared to be a Kamov Ka32. It had massive forks that could be fitted around the sawn down log; the log could then be lifted and flown to the logging base. The Russian crew were repairing the mechanism which operated the clamping forks with an old solenoid from a car and taping the electrical connections together. My confidence in the efficiency of this operation collapsed. In 2005 one of these machines crashed and burst into flames killing three of the Russian crew. As far as I know soon after this practice of logging was abandoned but in 1997 it was in full operation. My worries for the health of the catchment area of the Medalam River increased twenty fold.

The second was the raid on our camp by around 20 individuals from the Tabun Longhouse of Kuala Medalam. They arrived by longboat, a long way up the Medalam River seemingly without a purpose. They certainly had not arrived to say hello. They camped next to ours and proceeded to paint on the cave entrances in red paint the word 'Siga'. This was the name of their chief and it was obvious they were laying claim to ownership of the caves. Luckily they did not know where many of the caves were. I scolded one of their leaders as the paint would need to be removed at some time in the future as the intention was to establish a national park. Buda at that time was scheduled as a Protected Forest Estate and therefore came under the jurisdiction of the Director, as I had been appointed to supervise the expedition and collect all relevant information; I was officially acting as his deputy. The Medalam Protected Forest gave no privileges to the local community as it was classified as state land. They were therefore trespassing illegally without a permit from the Director. I was then fined for a sum of 300 ringgit and a parang (machete) with the threat of being hung from a tree with a caving rope. At a discussion it became obvious that all they wanted was to raid the caves for swiftlet bird nests. To keep the peace I paid up with the hope that they would soon disappear which they did. I could now see that forming a national park with the consent of the local community was going to be an uphill struggle.

I was now in a position to complete the documentation for national park status and submitted the 28 page document with the justification outlined. Being greedy I drew a map of the catchment area of the Medalam River of 50.7 square kilometres and submitted a proposal for the park to be extended bringing the total protected area to 113 square kilometres. You do not get anywhere without trying.

13

Gunung Mulu World Heritage 1999

'What lies behind us and what lies before us are tiny matters compared to what lies within us.'

Ralph Waldo Emerson

For many years I had considered that the Gunung Mulu National Park could achieve World Heritage status and a visit from some of the world's most eminent karst and cave scientists finally convinced me. Prof David Gillieson, Prof Derek Ford, Prof Elery Hamilton Smith and Prof Paul Williams came to Mulu for a look around. They all said the same thing that Mulu would qualify. I approached my boss who disagreed with concerns that Sarawak did not want foreign interference. I explained that this was not the case; the park would still be state-owned and controlled.

Years passed before Malaysia signed the World Heritage convention when a conference was called to discuss possible natural World Heritage areas of Malaysia that could qualify. Taman Negara in West Malaysia, Kota Kinabalu in Sabah and Gunung Mulu in Sarawak were suggested as likely candidates. I of course was asked to write a short proposal paper setting out why Mulu should be nominated; this had to include facts and figures, justification and integrity of the park. I jumped at the chance and, after a few months of work, submitted the documentation. I was not invited and my paper was presented at the conference in Kuala Lumpur, the capital of Malaysia by two high ranking officers from the National Parks and Wildlife Department; this appeared to be the normal procedure that the paper should be presented by a Malaysian. The findings were in favour of Kota Kinabalu and Mulu but there was a problem with Taman Negara as the park was situated within three different states without a common management policy. When I received a copy of the report on the conference the names of my two colleagues were credited with writing the proposal, again this appeared to be normal procedure.

In late 1998 I was asked to write the complex nomination for Mulu, a mammoth task which took me one year to complete. Justification was a problem as I wanted Mulu to be nominated on all four of the criteria for natural nomination which I thought it deserved.

The first is that Mulu 'contains superlative natural phenomena or areas of exceptional natural beauty and aesthetic importance.' This was not too difficult to justify. Number two was to prove that Mulu is an 'outstanding example representing major stages of earth's history, including the record of life, significant on-going geological processes in the development of land forms, or significant geomorphic or physiographic features.' This criterion was centred on the geology, geomorphology of the caves and cave fauna The third was even more difficult to justify, 'outstanding examples representing significant ongoing ecological and biological processes in the evolution and development of terrestrial, fresh water, coastal and marine ecosystems and communities of plants and animals.' The fourth was just as complex, 'contains the most important and significant natural habitats for in-situ conservation of biological diversity, including those containing threatened species of outstanding universal value from the point of view of science or conservation.' The integrity of the park was not too difficult as the National Parks and Nature Reserve Ordinance and the Wildlife Protection Ordinance all satisfied the criteria. The listing of all identified species for the fourth criteria was a head ache.

By 1999 it was becoming obvious that my time in Sarawak seemed to be of concern. The first was a poison pen letter written in Malay and sent to the government and signed by two of my Berawan relatives. From looking closely at the letter it was a photocopy and the signatures had been copied from another document and pasted onto the letter. They even spelt the name wrong of one of the persons who supposedly had signed. I obtained his IC and proved the letter was a forgery. I knew who had forged the letter but of course could not prove it. I never found out if my proof of forgery was ever officially accepted.

The second catastrophe was after 8 months my contract had not arrived. Without pay or expenses I managed to borrow a great deal of money from a kind tour operator and a contractor to survive as the wife and family needed to eat. This was not the fault of the Forest Department as they had applied for my contract as soon as the last contract expired. I had similar problems in the past of two or three months wait but 8 months was extreme. The head of the National Parks and Wildlife Department was Sapuan Hj Ahmad, a friend and a very hard working individual. We had an excellent working relationship and he accompanied me to the government head office where we inquired on why there was such a long delay. The answer was that the contract had not been signed by the State Secretary and after almost 10 years the Forest Department should have found a Malaysian to do the job. Sapuan knew that finding a Malaysian speleologist to work for government rates was not going to be easy.

Revival, Clearwater Cave, Mulu. Photo: Jerry Wooldridge

The breakthrough came when I was guiding the Minister of Tourism from Malaysia around Mulu and his assistant asked about my contract. I let it slip that 8 months was a long time to wait without pay, this was a private conversation so I was surprised it got to other ears. Within two weeks my contract arrived with my pay minus 30% income tax. I could now pay off my debts but was pulled over the coals by the Director of Forests for embarrassing the government. I should have kept my big mouth shut. It appeared that the Minister had mentioned it to the Sarawak Minister of Tourism informing him I was doing a great job but they should pay me.

I completed the World Heritage Nomination draught document and was summoned along with Sapuan to attend a meeting with the Minister of Tourism. Again I had a productive working relationship with the Minister as we had climbed up to the Mulu Pinnacles together some years previously and I had guided him around the caves. The Minister asked about the World Heritage proposal and was pleased to hear that the draught was completed. He asked about the possibilities of Niah National Park being nominated as a World Heritage Cultural Site. I explained about the quarry licence preventing this from happening. He then asked about a management plan for Mulu as required by UNESCO. Sapuan looked at me so I agreed to write it.

Betty and I were kindly invited to attend the annual conference of the Australasian Cave and Karst Management Association in Australia. This gave me a chance to hand over my draught of the World Heritage Nomination to David Gillieson and Elery Hamilton Smith for review. They did an excellent job with many suggestions. I was now in a position to recruit a team to help compile a management plan to World Heritage standards. David agreed and we worked on the Terms of Reference and budgets which came to less than 100,000 ringgit, approximately £20,000. I started work after completing and publishing the Gunung Mulu National Park Nomination for World Heritage Listing. I printed 20 copies for the Forest Department and government officials and posted copies to the Malaysian World Heritage Commission in Kuala Lumpur all within the designated period of time.

My contract had now expired and request for another one was refused. To get around this problem I was appointed by the Forest Department as a supervisor for a project I had recommended. This was electrification of the tourist caves via high voltage cables from park headquarters, thus eliminating the electrical generators at the cave entrances and the problems of carrying fuel. I was employed directly by the consultant engineering company based in Kuching. This enabled Sapuan to call on my services if or when required.

Just before my new appointment I received a phone call from Sapuan to drop the Mulu management plan as a contract for one million ringgit had been issued to the owners of the Mulu Hotel and could I attend a meeting at his office in Kuching. I was surprised considering I already had a team and had agreed to write the management plan for a very small sum. At the meeting the owner was there with a piece of paper which contained a long list of all the information that was required concerning Mulu for the management and development plan. I was commanded to supply this information and pass it on. The list was two pages long. It appeared that an Australian consultancy company had been awarded the contract to compile the Integrated Management and Development Plan but knew nothing about Mulu as they had never been there. I was now out of work so agreed to provide the information as I had little choice. An offer was made to pay me 5,000 ringgit for box after box of photocopied documents. Over the next few months I handed the boxes over to the manager of a hotel in Miri owned by the same company.

The World Heritage Nomination needs to be reviewed by many eminent scientists and evaluated by the International Union of Conservation and Nature, (IUCN). This is the official body nominated by UNESCO for all conservation matters including Natural World Heritage Nominations. Dr Jim Thorsell was appointed as the evaluator so I met him at Miri airport as it was my job to show him around and convince him that Gunung Mulu deserved the nomination. I took him to Mulu, guided him around the caves and gave an illustrated talk on the value of the park. Hiring a helicopter we flew around Mulu but he was disappointed in not seeing the pinnacles as they were covered in cloud. On leaving I asked him the million dollar question. The answer was positive and it began to look like one year of hard work could turn out to be a success.

A short while later I had dinner with the director of the Australian consultancy company when he described me 'as my own worst enemy.' The manager of the hotel where I handed in the boxes of documents smiling told me that 'I was considered as a very hot potato.' I soon received a phone call from the director of the consultancy company saying that my services were no longer required. When the management plan was published, there was strangely no mention in the acknowledgements of my months of effort.

Now I was not employed by the government meant that I had to give up my home which was government quarters and rent a house. Along with having to purchase furniture and pay for my car, I was once more broke.

The entrance of Lizard Cave. Photo: Peter Bosted

Sapuan asked me to compile the development and management plan for Gunung Buda as my paper on the national park proposal had been submitted to the government for approval. He had one million ringgit for the project. I completed the Terms of Reference and started the work but a short while later I received a phone call to drop the project as the allocation had been removed. I was then fined 1,000 ringgit by immigration as the company I was working for had not obtained my work permit. Problems seemed to be stacking up in ever increasing intensity.

It was not all doom and gloom as the next USA Gunung Buda Project was about to start and, considering my pay was derived from the Forest Department as a part of the contract, I was once more assigned to supervise the expedition.

14

USA Gunung Buda Project 2000

'You can't be brave if you've only had wonderful things happen to you.'

Mary Tyler Moore

This expedition was a little different as there was a specific scientific element of species identification, especially cave fauna which in the Buda caves was prolific. This would be essential information in order to develop a plan of management. The problem arose immediately as the Sarawak Biodiversity Centre, SBC had been formed with their ordinance claiming jurisdiction over all research in Sarawak. Of course in law this was not correct as Gunung Buda was a part of the Medalam Protected Forest so came under the jurisdiction of the Director of Forests. As the expedition had the Forest Department permit and a signed Memorandum of Understanding, a permit from the SBC was not necessary. The department insisted that an SBC permit should be applied for just to keep the peace so we did. The USA team applied as instructed and I went to the SBC office and handed over the applications three months in advance of the expedition start date.

This time they had 30 team members but as I was no longer employed by the department all I could muster were two Mulu guides to join me, Roland Gau and Johnny Baie Hassan, both were from the Limbang District and excellent in caves and the forest. Once more Tropical Adventure Tour Company arranged all the logistics with tremendous help from Chua Eng Hin. Chua lived in Limbang and had been assisting on the Buda expeditions as regards to transportation and food supplies.

The first and foremost objective was to attempt to find a connection from Green Cathedral to Snail Shell Cave or Turtle Cave as they were so close together this would give us a combined length of over 37 kilometres. The big discovery was the Babi Ballroom, a magnificently decorated chamber. They found this via yet another entrance just below the main Green Cathedral doline. The various entrances to this system were mounting up. Connection fever had set in and a new entrance to Snail Shell was found emerging way up in the mountain. They called it the Long Way Home as the surface traverse down the karst back to camp was arduous to put it mildly. Upper Turtle Cave received much attention as another climbing lead led to an entrance into the Lost World Doline discovered during the 1997 expedition. This entrance called Coiled Snake enabled a detailed traverse of this doline through horrendous pinnacle karst where Dragon Cave was discovered. It was hoped that this would connect to Snail Shell but disappointedly after 224 metres led to another entrance with no connection found. Yet another promising looking entrance was explored which they called Longhouse Cave but again it only led back into Upper Turtle after 138 metres. Attempts to connect Loris Cave to the Turtle Cave River failed being only 10 metres apart. These connections were proving to be very elusive.

Green Cathedral showed its ferocity when Vivian Loftin on a climb down fell 11 metres when the handholds crumbled. With luck she landed in a sump pool of water about one metre in depth along with a lot of rock. Covered in cuts and bruises but with no broken bones her companions rigged a rope so she could climb out. The last trip into the system was in March when Joel Despain, Jean Krecja and Mark Fritzke looked at Jeffrey's Passage The ropes required to pass beyond that had been left in the cave in 1997 and had disappeared, probably stolen by bird nest poachers so they retraced their steps in order to bypass the obstacle to the Sink City entrances. Again with no rope to pass a traverse they retired and were joined by Kate Lysaght the following day. Mark pushed through a tight vertical fissure with a strong air flow, he returned to report a much larger passage beyond. This remains as a good going lead for another year. The expedition mapping increased the length for 4.48 kilometres bringing the total length of Green Cathedral to 25.7 kilometres.

Ant Farm Cave was developed at base level with a small stream to a downstream entrance, mapped for 117 metres. This type of cave appears to be common in tropical areas where the steep cliffs are eroded by alluvial streams with high carbonic acid concentrations. Good examples can be found in Mulu and Niah. Water Works Cave was another example mapped for 648 metres with a hoped-for connection to Green Cathedral via high level leads.

Coffin Cave was carefully mapped and investigated by Jean but the conclusion is that this cave should be classified as out of bounds due to its archaeological significance.

Herb Laeger's work on the Snail Shell ramps in 1995 and 1997 was spectacular at reaching high up in the mountain at 470 metres above the entrance, the highest or deepest cave in Malaysia. He was not done with Snail Shell and returned this time armed with a rock drill. Many further steeply-sloping ramps were climbed but, although a strong air current was present, the ramps terminated in passages too small for humans. Again some of the ropes left in place by the last expedition had disappeared; the bird nesters must now have a good collection.

The Pig Sty Pit had been descended by the 1996 team to masses of mud and a sump. A piece of blue flagging tape had been attached to a rock at the furthest point reached. Herb and his team noticed an air flow at the top of the pitch so decided in their wisdom to take another look. The pit was 30 metres deep and the sump was not there anymore. The mud on the walls extended above head height and they found the piece of flagging tape 3 metres above the muddy sump pool which was where the air was coming from. They called the sump the Plumbers' Trap as it obviously filled to the roof when it rained. Beyond was glutinous mud but the passage was larger so they mapped onwards arriving, after about one hour, at another 45 degree clean-washed ramp with a small stream cascading down. Herb and Pete Shifflet climbed and while Herb waited Pete continued upwards for around 60 metres to a section which required bolting for safety. That night it rained hard so they gave it six days before returning to the Plumbers' Trap Ramp but there was no airflow and the sump was completely flooded. Any exploration in this area is fraught with extreme danger as it would not take long for the sump to fill with water trapping any explorers on the far side. This lead still remains, is this the long sought-connection between Snail Shell and Green Cathedral?

From the base of the massive extensive ramps a small passage led off surprisingly into a well decorated large chamber. Three pits in the floor led down to a clean-washed river passage, the downstream continuation of the Snail Shell River found in 1996. This had been dye tested to the Turtle Cave River so, in very dry conditions; maybe it could be followed to the high level continuations which seem to be normal for the Buda caves. The strange thing about Snail Shell is the profusion of these ramps, not found in any other caves either in Mulu or Buda. There does not appear to be a reason why, as these ramps normally lead upwards to large horizontal passages, the older levels of cave development. The total Snail Shell length came to 11.9 kilometres.

No one had been beyond North Gorge so a decision was made to build a small bivouac camp where exploration could be made without the very long hike back to Base Camp in the Medalam Gorge. This was affectionately named Happy Camp. The floor was built of sticks raised a little off the ground so to me was akin to sleeping on a bed of nails. The following morning I had an ear full of blood as the leeches had been having a wonderful time.

Catfish River Cave was at base level with a small stream mapped for 591 metres but full of Catfish. A high level lead found a boot print. Had this been left by the British Cavers in 1980 as Compendium Cave was supposed to be close by? Roger Mortimer and I climbed the hill to the north and, via the smell of bat guano, found a nice well-decorated cave terminating in flowstone with a small hole blowing air. We could see beyond into what looked like a large tunnel.

The following day I had to get back to Base Camp as I had a meeting scheduled in Kuching with my employer and I needed to find out why our SBC permit for research had still not been issued after three months. It was a long lonely walk back taking a few hours.

Compendium Cave was found as a 30 metre by 30 metre entrance. The British team had mapped 1.9 kilometres with many unexplored leads so, to make it easier, the USA team started the mapping from scratch. The leads marked by the British as draughting produced a good going lead heading north. This terminated into a pitch down to known passage but to the south a fine tunnel led to a small draughting hole. They forced a way through and emerged in the cave Roger and I had explored.

Surface searching across the karst proved to be severe when described as the most dangerous terrain ever seen, one slight slip being fatal. 10 metre high pinnacles with 30 metre deep holes covered in rotting vegetation, it took one hour to travel 30 metres.

Later Mike and Andrea Futrell along with Steve Smith entered the now named Assam Cave via the Viper's Lair entrance to an old bird nesters pole which disintegrated when they tried to climb it. Returning with ropes they lassoed some projections for aid and climbed up to a large tunnel with good air flow. They halted in a large room with a drop down requiring a rope. On the last day of the expedition they descended the hole to a large passage with hundreds of leeches hanging in the roof, not a place to linger. After a few hundred metres they found another entrance but could hear radio music. Looking out down below in the jungle was the camp of the bird nesters. They mapped onwards in a large passage passing entrances and skylights. The walls were marked with graffiti, Tubun Caves indicating their claimed owner ship. By midnight they had mapped a further 1.3 kilometres with no end in sight and many leads everywhere. Running out of water and food they headed back out. The system was heading directly towards Hornbill Cave. Over the course of the expedition they mapped the cave to 4.18 kilometres. To save confusion it was given the name, Assam Caves.

The Hearth, Babylon Cave: Photo Dave Bunnell

Hornbill Cave did not disappoint with a streamway being mapped heading west towards a large resurgence on the west side of the mountain. This resurgence was entered to a waterfall with crumbling rock and no airflow so was left for future cave explorers. Hornbill had many marks on the walls indicating once more that this cave was being raided by bird nest poachers. They certainly seemed to get everywhere. The cave finished only 300 metres away from the Assam Caves so a connection does seem likely. 1.9 kilometres were mapped bringing the total to 3.4 kilometres.

The source of the Assam River was found around the doline where Happy Camp had been constructed with a number of short river caves but the terrain was difficult to negotiate.

One major objective was the connection of Babylon Cave to the British discovery of Disappointment Cave where all leads had petered out. The two caves were close by so a connection did seem likely. They were not disappointed bringing the total length of this highly complex cave to 5.2 kilometres in length.

Spirits' River Cave was described by the USA team as the Big Cave in the Wrong Place and quite rightly so, as it has no right to be there. There does not appear to be any limestone as all the surrounding land is shale and gravel. The entrance was found by the driver of our vehicle and was pointed out to the team. The problem was that it is situated outside of the designated boundary we had drawn for the proposed national park. Just off a logging road is a steep-sided heavily vegetated valley with a small stream running into the valley from the south with limestone at the base. The cave was ridiculous as a complex series of interconnecting passages with sky lights into beautiful tropical areas. The entrance led to a 40 metre high canyon stream passage with a rock bridge. 15 to 30 metre diameter passages led from one collapsed doline to another all containing jungle. Many of the passages were at high level indicating various stages of development. The limestone does not appear to be more than 60 metres in thickness but its extent is unknown as we have no idea how far it extends beneath the shale cover. Some pottery and bones were found which indicates it has been known for many years by the local community and possibly used as a burial site. There are many leads left to explore and map so the true extent of this strange cave is not known. The mapped length came to 5.2 kilometres.

The Buda River Cave came as a surprise being so close to camp as limestone could not be seen. It is situated in an isolated block to the west of the main southern block but surrounded by jungle. The main Buda

River, which issues from Turtle Cave and Palm Beach Rising, flows through the cave so a through trip can be made, most of the time floating on rubber tubes. The trip finishes in a beautiful river canyon with 8 metre high sides. With high dry levels overlooking pinnacles and the canopy of the surrounding jungle, this is a perfect tourist trip where visitors could be kitted out with helmets, lights and buoyancy aids. Including the Little Buda River Cave and the high level passages, over 2 kilometres were mapped with numerous entrances.

The team could not keep away from Deliverance Cave, situated on the north side of Gunung Benarat in the Gunung Mulu National Park. Multi day exploration was conducted by Chris Andrews and Mark Rosbrook with a bivouac just inside the entrance. The first objective was the high level lead known as Coral Dome. With a battery-powered drill and a mountain of climbing gear they climbed 53 metres in three pitches ever upwards placing runners and bolts in the wall over a period of a few days. Mark descended a 22 metre pit using the remaining charge left in the drill in order to fix a bolt and hanger in the wall. They returned back to base camp to recharge the battery in the drill. Leads still remain in the Coral Dome which appears to be draughting.

The two returned to their bivouac in the cave a few days later but this time decided to progress to the far southern end of the cave. One climbing lead looked far too dangerous to tackle with loose rubble everywhere but Mark was not put off and grovelled around the knee deep guano looking for an air current between the massive stalagmites. He succeeded in finding a small blowing passage near a 5 metre high stalagmite and the pair began mapping through "rooms" to a number of pits in the floor. The next day they returned and descended a 33 metre pitch to explore incredibly decorated passages with helictites, anthodites and aragonite crystals in profusion. They called it Euphoria and the cave was extended in length to 4.3 kilometres.

Only 2 kilometres divides the southern end of Deliverance to the northern end of Benarat Caverns. The Mulu dream has been a complete exploration of the Clearwater Cave system from the Melinau Gorge to the resurgence in the south and the same applied to Gunung Benarat, a complete underground traverse from north to south is still a distinct possibility.

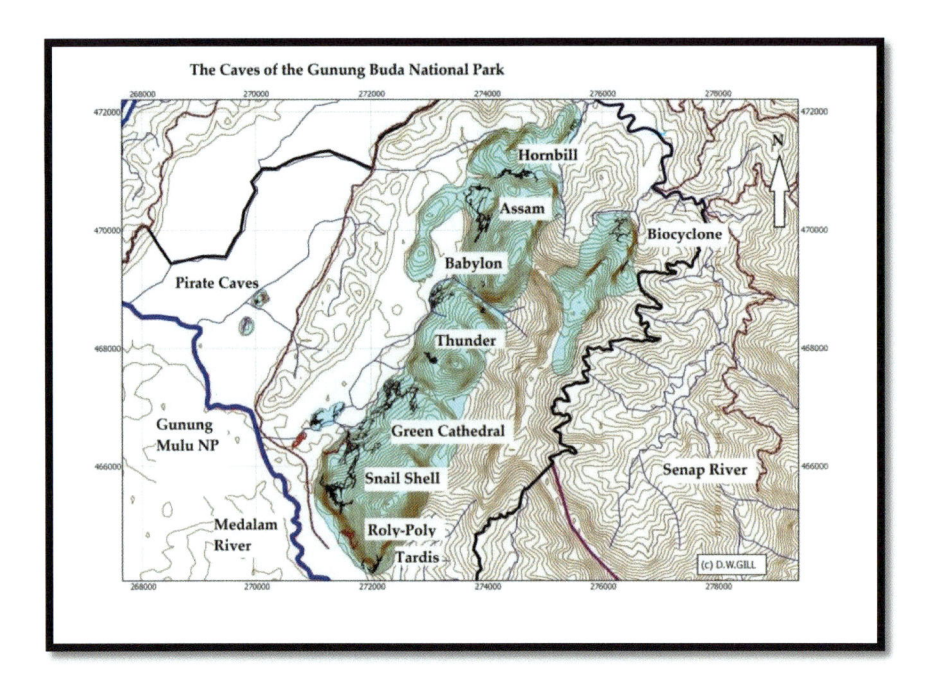

The cave systems of the Gunung Buda National Park, Sarawak, Malaysia

The invite to attend a party at the Kuala Medalam Longhouse was accepted with traditional dancing, lots of food, tuak (home-brewed rice wine), speeches and hangovers the next morning. Again it became obvious that the main interest was to claim Buda as traditional land so the caves could be raided for the valuable bird nests.

Later in the year Buda was constituted as a national park and no privileges were granted to the local community. This means that all entry into the area requires a permit issued by the Forest Department via the National Parks and Wildlife Department-which includes the local community. As regards the process of conservation, this was a move in the right direction but illegal nesting could not be stopped without vigorous enforcement.

A very sad part of this splendid expedition was that Ronald Wayne Simmons died for a second time while mapping during a solo dive in Allan Millpond Cave, Florida on the 14th February 2007. He had completed 43 dives in this flooded maze cave. I call it the second time as the first time was when he almost drowned in the Medalam River with Herb Laeger on a previous Buda expedition. Ron was a great cave explorer and a fine man.

Another catastrophe occurred at the end of the expedition. I had a satellite phone which was as heavy and large as a building block; communication was possible with great difficulty from one spot just outside base camp. I had managed to meet the director of the Sarawak Biodiversity Centre during my visit to Kuching. He assured me that the research permit had been signed by 30 people, only the Director of Forests and the State Secretary had not signed as they were in Australia but would sign on their return in two days time. At base camp I received a message passed on via the jungle drums that the SBC were on their way with an armed escort. All the team were arrested and all the scientific research collection confiscated. As this area was the Medalam Protected Forest it came under the Forest Ordinance 1958-which stipulates that the Director of Forests has complete jurisdiction over the area designated. As we had a permit from the Director and a Memorandum of Understanding as a combined Sarawak and USA expedition, no laws were being broken. The SBC Ordinance was dated 1997 so did not, under the law, supersede the Forest Ordinance. The SBC refused to listen to this argument and Special Branch of the police department advised me to take it to court which was impossible as the USA team had jobs to go back to. The USA team were fined a substantial amount and they sent an email to various people explaining the events. I sent this to the Forest Department as I was obliged to do, but, for some unknown reason, it was sent to my out of date email address list. This could have been via a virus or my mistake as normally; if an email is sent to an address list it is Bcc, which means the email addresses are not shown. Events started to mount up with ever increasing difficulty.

Our four expeditions had clocked up 92 kilometres of explored and mapped caves, a remarkable achievement by a fine group of dedicated cave explorers.

I was of course delighted that Gunung Buda was constituted as a National Park and I considered it as a major achievement over a period of eight years of work.

15

Goodbye Sarawak

'Pain is temporary. Quitting lasts forever.'

Lance Armstrong, Sally Jenkins

I was back once more in supervising the contract at Mulu for the electrical distribution of power cabling to the caves. As I had nowhere to live, I slept on the floor in one of the spare buildings. The job was very boring as the contractor knew exactly what and how to do the work.

At a conference on World Heritage at the hotel in Mulu I was advised to keep a low profile as certain people were upset about the email from the USA team that I had forwarded to the Forest Department. I was not asked to give a talk on Mulu as Sapuan did a good job with an illustrated lecture. Advice came via the Minister to email an apology for sending the email which I did-to no avail. A phone call from the Miri Director of Immigration demanded my immediate return to Miri. I called in his office along with Betty, Jeff and Racheal and was told to leave Sarawak and show him my flight tickets. No reason was given as this instruction had arrived on his desk from a higher authority and he could not show it to me. I told him I would go to Brunei which he would not accept but as it's an independent country there was nothing much he could do about it.

My only choice was to escape to Brunei for a day and return in order to obtain a tourist four week stamp on my passport. This worked for a while until someone informed immigration what I was up to and I was refused entry. I did show them a copy of the letter from a high ranking government minister explaining that he was making attempts to have the decision reversed. They photocopied it but still refused entry back into Sarawak. I was left abandoned with a few hundred ringgit and the clothes I was wearing at the time but managed with help to find a friend where I could live for a while. Betty came every week with news, clothes and money bringing along Jeff and Racheal.

As alcohol is banned in Brunei at weekends we went to a café where we could drink beer from a tea pot into tea cups. The place where I was staying was full of Berawan relatives as the kind lady in charge was

Betty's cousin and managed a string of phone shops. One favourite pastime for foreigners in Brunei is to drive across the border into Limbang Sarawak and spend a night drinking beer in a wooden hut. We tried it and immigration let me through with no problem.

My four year old daughter Racheal came to stay with me for a while and one day I told her we were going for an adventure. With a small bag we caught a boat from the jetty which took us through the mangrove forest along the estuary of the Limbang River. In Limbang Sarawak we had no trouble with immigration as Brunei employs many foreign individuals involved in the oil business. Stepping out onto the pavement of the busy road a truck pulled up with the shout to get inside fast as immigration had warned the driver not to assist me. He deposited us in a small hotel and next day undercover transported us to the airport. A domestic flight later and we arrived in Miri to be bundled into a car and sped away fast.

I was driving along the main road one day when Betty said that a man was waving to me from the back window of a limousine car. I looked to see a high prominent government minister waving and giving me the thumbs up. I am sure he knew I was not supposed to be in Sarawak.

Now, after many months of trying, it became apparent that I could not obtain a work permit and that my only alternative was to dispose of all our belongings, house, furniture and car and go back to the UK. Drawing from the bank what little money we had left, I changed it into pounds and dollars and headed back into Brunei with the family. Immigration was surprised to see me back once more in Sarawak. It was the best idea under the circumstances, as also my dear parents were nearing 90 years of age and had expressed concerns that they did not want to be incarcerated in an old people's home-but would prefer to die in the home they loved. As the only child it was my duty to do everything possible for them so leaving Sarawak at this time was a necessity.

Via a friend I obtained a personal interview with the British High Commissioner in Brunei and explained the situation. It appeared that my friend and the High Commissioner had been to the same college in the UK, probably Eton. There was no problem in obtaining a British Passport for Racheal as I had her birth certificate registered at the UK embassy in Kuala Lumpur Malaysia. I had guided the British High Commissioner for Malaysia around Mulu a few years back. It seemed like it's not what you know but who you know, or the old school tie syndrome. I also obtained a long term indefinite permit to remain in the UK for Betty and Jeff stamped in their Malaysian passports. This had been a long and hard struggle but we were now set to leave.

I arrived back in Manchester in January 2001 with my wife, two children and a suitcase and still broke-but the past 10 years had been an exciting and wonderful experience. The good news was that I had achieved success with Gunung Mulu National Park being nominated for World Heritage status on all four of the criteria and Gunung Buda being proclaimed as a national park. This represented 10 years of effort collecting, analysing and formatting all data obtainable for Gunung Mulu in order to prove its Universal Significance and to outline the justification. Thanks was never expected or received.

Some would call it traumatic leaving Sarawak but to us it was just life's little ups and downs although it did seem strange to be back in the same house where I had lived for over 22 years since 1946. It seemed I had come around in a full circle with Jeff at my old senior school and Racheal at my old junior school but the main problem was to find a job. The Job Centre was a waste of time as they failed to understand what a speleologist was or a national parks and wildlife officer. They tried to get me a job as a town park gardener or going back to college to obtain a certificate for electrical engineering. At the age of 60 years I considered this to be out of the question so became an odd job man. My beautiful little cottage in Chinley Derbyshire had been sold and my first son Mark obtained a house of his own with my help. We survived thanks to mother and father as dad bought me a second hand car so we had wheels. I had a hard time laying flags and block paving with an Irish friend as it damn near killed me. Turning up at a house inhabited by a young lady who was expecting to see two young muscular men with tight jeans and showing their bum cracks when they bent over, she was a little surprised to see two old age pensioners laying flags.

When my parents celebrated their 60th wedding anniversary they received their rubber-stamped signed card from the Queen and a visit from the Mayor of Tameside. When asked by the local press what their secret of a long marriage was, they replied that they always settled an argument before retiring to bed.

Betty was superb at looking after my parents but the saddest day came when father was taken into hospital. I just hoped he would last a couple more years but I did thankfully get eighteen months before he passed away. He served his country during the war, his church as a Sunday School Teacher, his community as a volunteer teacher at the Church of the Resurrection School in Beswick where he taught children having reading difficulties and he certainly served his family. I was chatting to some ladies once in a pub in Droylsden and one asked me if my name was Frank Gill as I must have looked a bit like him. I said no, but he was my father. She asked how he was and when I told her he was dead, she burst into tears. She explained that she was a manageress in a clothes shop and if it wasn't for him she would never have a job as he taught her to read. Mother gave up and died one year later so I buried their ashes under an oak tree in the Memorial Forest in Longdendale, Derbyshire. When in England I always try to go and see the acorns.

I still have the same house which has been in the family now for 74 years.

Dad somewhere in France 1944 with his guard dog. Photo: unknown

16

French World Heritage Caves Nomination

'Life is too short to think small.'

Jonathan Murray

Of course my interest was still centred on caves and conservation rather than flagging. I had kept in touch with Dr Jim Thorsell of the International Union of Conservation of Nature (IUCN) and he asked if I would accompany Prof John Gunn to evaluate a World Heritage Nomination in March of 2001. I of course jumped at the chance; there was no pay just our expenses but a great opportunity to see some of France's greatest caves which are normally out of bounds for cavers. The formations within these caves are delicate and rare so access to these important cave sites is strictly controlled, most having huge impenetrable locked steel gates.

The nomination was named 'The Ensemble de grottes a concretions du sud de la France' and consisted of 18 separate cave sites with a total of 24 different caves; the most spectacular and beautiful decorated caves in France. A total of 135 kilometres of cave passages was included in the proposal.

Early on in the year John Gunn was asked to attend an IUCN meeting in Paris to discuss a proposal for World Heritage status for these magnificent cave systems. A number of external evaluators had raised serious misgivings on the nomination, as the proposal was a scattered serial site with little common legislative protection, management plans or boundaries. The external evaluators had not visited the sites, so assessments had been made purely on the documentation supplied. The nomination documentation was written entirely in the French language and was based only on the speleothems and rare minerals found, mainly aragonites. The proposal claimed that the caves satisfied the UNESCO World Heritage natural criteria which states that, 'the site must be an outstanding example of on-going geological processes with significant geomorphic features' and 'be an area of exceptional natural beauty of aesthetic importance.' The integrity of the sites must also be satisfied as regards to adequate legal protection, conservation and management. This applies not only for the known cave itself but also for the terrain above the extent of known and projected cave passages. All the underground and surface catchment areas should be included within the protected boundary. There was at that time some doubt as to the ownership, protection, management and the conservation measures applying for each particular cave.

It appeared that out of an original list of over 40 separate cave sites proposed by a 30 strong team of French speleologists, the final 18 had been chosen after much debate as representative of the finest decorated caves in France. The proposed caves contained the rarest speleothems known in the country if not in the world. They differed somewhat from other cave World Heritage sites throughout the world and especially those containing rare minerals. Meteoric waters were responsible for the formation of the majority of the French nominated caves. World Heritage Carlsbad Caverns area in the USA, which also includes Lechuguilla cave, probably the most beautifully decorated cave in the world, is also known to be of hydrothermal origin.

In Paris John was presented with an outstanding collection of slides by one of the authors of the World Heritage document, Patrick Cabrol and arrangements were made for a site visit. It was planned to visit 12 of the 24 nominated caves in 9 days, a tight and hectic schedule as the sites were widely scattered around the Pyrenees, Montagne Noire, Montpelieraine, Grand Causses, Ardechois and the Alps. It was impossible to visit all the caves in the allotted time as some were situated in high mountain regions, still snow covered at that time of the year. As the criteria are open to individual interpretation it was decided by IUCN to send two evaluators, John and me, for the 9-day trip.

On 23rd March Patrick Cabrol met us at Toulouse airport. Patrick was to accompany us throughout the mission and had arranged all the logistics. The first day was spent in the two kilometre long Gouffre d'Esparros in the Pyrenees. This cave had been made famous by Norbert Casteret in one of his books and contains outstanding massed pure white aragonite needles in the lower reaches. A tunnel had been driven into the lower passages and the cave partially opened to the public. Access beyond the show cave into the delicate areas is restricted to speleologists only and is barred by a huge steel gate. A marked trail was followed along this amazing crystal lined passage avoiding touching the walls. The cave received legislative classification in 1987 under the French 1930 Environmental Code and environmental monitoring is in progress in the show cave sections. The same classification also applied to some of the other caves on the list.

The author in one of the proposed French caves. Photo: John Gunn.

The second day was centred in the Montagne Noire region with a visit to the village of Lastours. This is the site of the famous Grotte Bleue, also referred to as Barrencs de Fournes with Roman mine workings above. This is not actually a cave but a mine listed in Hill and Forte book of 1997, *'Cave Minerals of the World'*, as one of the top ten mineralogical caves. A substantial wall seals the entrance, as the blue aragonites are delicate in the extreme with some unique forms. The evaluators argued the case that although there are many photographs of the speleothems a site visit was justified as the mine had been sealed since 1974. One worrying aspect was the nearby gold and arsenic mine and the lack of legal protection but plans were in hand to re-open the Blue Cave in a few years time along with legislation in order to fully protect the site.

The third site visited was the 2-kilometre long show cave of Limousis containing a huge hanging aragonite cluster weighing 9 ton, probably the largest single cluster known. The cave was first opened to the public as long ago as 1932 but was well known in the early 1800's. Unfortunately much damage was done to the entrance sections of the cave by Italian miners due to calcite extraction in and around 1820. Limousis is one of five caves within a single site, which includes Trassanel, Gaubeille, Embuc and Cabrespine. Gaubeille and Embuc contain masses of aragonites while Trassanel is well known for its large 'calcite disks' a plate formation growing from the walls. Time did not permit a site visit to these three caves but the show cave section of the 18 kilometre long Cabrespine was briefly visited. The associated river passage of Cabrespine includes all forms of speleothems known. A common boundary is planned in order to protect all the caves within this site so at that time both sites did not satisfy the criteria of integrity.

The third day on the road saw the evaluators back in the Pyrenees. This time it was the turn of the TM71. A Nature Reserve dates from 1987 and covers all of the 9.5 kilometres of this superb cave system, the entrance being protected with three substantial steel gates wired with alarms via the telephone lines. The evaluators were suitably impressed with the six hour trip ably guided by Philippe Moreno, a professional wild cave guide who made sure we knew how to use our supplied equipment. The cave contains a spectacular blue aragonite cluster, strange red calcite blisters, pure white monocrystalline triangular stalagmites and much more. Aguzou on the opposite side of the gorge is equally as spectacular and is also fully protected as a classified site. Unfortunately time did not permit a visit to this cave but Philip organises guided tours for speleologists during the summer season.

Day 4 on the 26th March the evaluators had the unique opportunity to visit the famous Lachambre, a cave we had often read about. This 25 kilometre-long system contains kilometres of mind-blowing white aragonite-encrusted passages from floor to ceiling, cave pearls, hydromagnesite formations associated with magnesium and a unique vein of talc, which crosses the cave passage. A truly unique experience to wander for kilometres along high passages completely lined with pure white aragonite flowers. Great care was taken to make sure we did not touch the walls of delicate crystals. Of all the caves visited the Aladdin's cave of Lachambre stands in its own right as a cave worthy of World Heritage status. The site was classified in 1991 and is well protected via a bombproof steel gate at the entrance.

The same day we made a mad dash to the Grand Causses and a private evening tour of that well-known show cave the Aven Armand, first descended by the father of speleology, Edward Martel. This classified protected cave was first opened to the public in 1927 and the main chamber can now be reached via a tunnel and a railway. The sheer numbers of tall stalagmites are worthy of note so the cave could be regarded as being representative as an integral part of the nomination.

The first cave to be visited the next day was a disappointment coming so soon after the delights of Lachambre. Demoiselles in the Region Montpelieraine is another well known and frequently visited show cave. The original tunnel entrance and tramway dates back to 1929 and the pathways from 1930 onwards but the cave has been visited on a regular basis from the early 1800's. The massive calcite draperies and huge 20-meter high stalagmites are encrusted with layers of soot from the torches used. Calcite is everywhere in the cave and dating reached the limit of uranium series at 350 thousand years before present. This cave has been included as one of the sites for its large and ancient speleothems.

Clamouse was next on the list, another fine 6 kilometre cave first explored in 1945 and opened to the public in 1964. Clamouse exhibits fantastic aragonite, hydromagnasite and calcite formations protected with glass panels. Calcite formations include a 7 metre long stalactite and a very large disk plate formation.

Days 6 and 7 saw two fairly exhausted evaluators back in the Montagne Noire region descending a well-kept secret called the Asperge. This 8 kilometre cave along with PN77 at 5 kilometres in length and Grotte du Rautely at 1.5 kilometres have been grouped together as one of the sites as they are all part of the same hydrological system. The karst is rugged with wild boar roaming the woods and full protection in the offing as a Nature Reserve. The caves are pretty rugged too. Asperge is protected with two steel gates, one electrically operated and the cave has been dug and blasted in many parts. Numerous ladder descents, climbs and crawls are finally rewarded with an amazing blue aragonite needle cluster and a pale blue aragonite stalagmite, while the equally protected PN77 contains strange brown and white-tipped aragonite straws not reported from elsewhere in the world. The speleologists concerned in the protection and opening-up of these caves deserve every accolade in their efforts of conservation.

The mission was fast drawing to a close but not before a lengthy 9-hour trip into the far reaches of the classified 5 kilometre long L'Aven Orgnac in the Ardechois region. Orgnac is one of Robert de Joly's original explorations and his heart is still there, placed in a casket in the show cave. The show cave was hardly looked at in the rush to get into the big passages. The evaluators began to make comparisons with the caves of China and Sarawak as regards to the sheer size of the tunnels and the volume of 20-metre high stalagmites. Masses of helictites covered the ceilings and we followed the huge tunnel down to large calcite crystal lined gour pools and passages with numerous shield like formations. Being one of the finest decorated show caves in France; the French government have placed great importance on this tourist attraction by investing three million euro's in driving a new access tunnel. A build up of carbon monoxide levels was of some concern but has now been alleviated by removing one of the car parks from over the top of the cave and constant monitoring is in progress.

The final day consisted of a quick trip to the playground of British cavers, the Vercors Region in the Alps. The Choranche show cave with its 3-meter long soda straws and light show is well known and well protected. Although the area is a Nature Reserve, proposed classification will ensure long-term conservation. The classification will include the complete system of the Gournier at 20 kilometres in length and the nearby Coufin at 13 kilometres in length along with its fine travertine falls.

All that remained was a fast drive to Lyon to catch the plane home, a punch up with a rude tourist and to write the evaluation report.

The caves we were unable to visit include the Grotte de la Cigalere in the Pyrenees. As I had previously visited this system in 1976 and it is also extremely well documented in Sid Perou's film it was not considered necessary for a site visit. Cigalere is a classified site and has been included for its rare gypsum speleothems, which include, black gypsum, iron hydroxide and manganese hydroxide minerals and the largest gypsum flowers in the world. Only one expedition per year is allowed into this system at the present time and a huge gate protects the entrance.

Mont Marcou in the Montagne Noire region contains rare green and yellow aragonite formations but at that time was not a classified site.

In the same region Pousseliere is packed with aragonites but still requires classification, which has been proposed. This cave is protected with a substantial gate and controlled access via the owner and the local Speleo Club. Grotte de Lauzinas has been proposed because of its strange and unique mud formations, which include one metre high mud mushrooms, mud fir trees and also cave blisters filled with gypsum, opal or sediment. In the Grands Causses two other caves had been proposed, the Aven des Perles and the Amelineau, both still need evaluation at a future date.

The general impression after the site visit was that these caves deserve World Heritage status but with a number of provisions before they are considered for inclusion on the World Heritage list. All the sites should have a common management strategy, legal protection including catchment areas and also further evaluation needed to be undertaken on the sites not visited.

We wrote and submitted our report to IUCN hoping to be asked to look at the cave sites we had not had time to evaluate. It is still on the World Heritage Tentative List and should eventually be scheduled as these caves certainly deserve to be included being some of the most beautiful and unique caves in the world.

17

A few cave and karst projects

'When you reach the end of your rope, tie a knot in it and hang on.'
Franklin D. Roosevelt

During my isolation in Brunei I stayed at a friend's house, an oil engineer working for Brunei Shell. He was from the Sultanate of Oman and not having much fun in Brunei travelled to Miri most weekends for fun and games. Eventually he was transferred back to his home country and invited me over as his friend was involved with the Ministry of Tourism who intended to develop Al Hoota Cave as a tourist attraction. Being involved with cave development for tourists for so many years in Sarawak it sounded like a good idea. Becoming bored I decided on a visit but needed to do my research first as it is important to know what you are getting into as I knew little about Oman.

Oman had a chequered history being an important trading hub and came under British influence in the 20th century. As an absolute monarchy it was ruled by Sultan Said bin Taimur from 1932 to 1970 with little development of the country implemented. There were few roads and only one hospital and school in the whole country. His son, Qaboos bin al Said was a different matter, educated from the age of 16 at Bury St Edmunds UK he attended Sandhurst and became an officer in the British Army. On returning to Muscat and Oman-as it was then called in 1966 -, his father put him under house arrest. A bloodless coup was arranged by MI6 and British Armed Forces which was sanctioned by Prime Minister Harold Wilson in 1970 and Qaboos became the Sultan. He arranged for massive development of his country and invited Oman nationals to return to help with development. He changed the name to Sultanate of Oman. My friend was from Zanzibar the one time hub of the slave trade which had been a part of Oman territory. Believe it or not, slavery in Oman was finally abolished by the Sultan in 1970.

The main problem at that time was the pro-Soviet led Dhofar Rebellion from Yemen which went on for 10 years from 1965 to 1975. With the help of British Forces the rebellion was finally crushed. Hearts and minds of the community had played a special role. The Battle of Mirbat is worth mentioning as nine SAS soldiers with a few soldiers from Pakistan and Oman were attacked by a force of 250. Fijian Sergeant Talaiasi Labalaba was killed in action while operating a 25 pounder artillery piece on his own which is usually manned by six people. Trooper Thomas Tobin was also severely wounded and later died. The rebels withdrew with heavy losses but the battle was played down by the British Government although it is now a part of SAS legend.

The Sultan passed away in January 2020 and, being childless, was succeeded by his first cousin Sultan Haitham bin Tariq al Said who had also been educated in the UK at Pembroke College Oxford.

As for Al Hoota Cave this was situated at Al Hamra at the foot of the Jabal Shams Mountain which rise to an impressive 3000 metres. The cave had been mapped for 4.5 kilometres with an outstanding 800 metre long and 10 metre wide lake. The lake contained a blind cave fish *Garra barreimaie* so required conservation measures.

Dave Clucas fancied a trip so we flew to Abu Dhabi in the United Arab Emirates and were impressed with the cosmopolitan atmosphere as 80% of the population of 9.8 million in this small country are foreigners, mostly Indian and Pakistani. There is a definite shortage of females as only 28% of the population are women. We met my friend and drove through the desert to Muscat the capital of Oman. Arranging a meeting with the Ministry of Tourism was not a problem but sadly it appeared a contract had already been issued to an Austrian company for the project planning of the cave development-but we obtained permission to explore Al Hoota Cave. The limestone appears to be Late Triassic overlain with Jurassic pelagic limestone but a very arid landscape with not a tree in sight. After spending 10 years living in a jungle I found it stark. We struggled to reach the entrance in the heat but the cave was large and hot with a few old inactive formations. The lake was impressive and thankfully this was to be left as a part of the natural cave and not to be developed for tourist trips. In this arid landscape any heavy rain results in massive run off resulting in floods and it seemed to us that this sink to resurgence cave could flood on the onset of rain, maybe once every 10 years or so.

Al Hoota entrance with the tourist access tunnel on the left under construction. Photo: David Gill

The desert and mountain ranges of Oman are impressive with massive canyons, some are 1000 metres deep. We borrowed a 4-wheel drive and explored one of these massive canyons looking for caves, we only succeeded in getting trapped in boulders with not a cave to be seen. The springs are spectacular and along with the World Heritage clay forts are well worth a visit. IUCN asked me to investigate the problems with the World Heritage Arabian Oryx Sanctuary so I arranged an interview with the director of conservation. It appeared that this rare and beautiful creature was illegally hunted and sold to private collectors for huge sums. Little could be done to protect them without an army of rangers. I submitted my report to IUCN and, sad to say, this World Heritage Area was placed on the endangered list and in 2007 removed as it had lost its Outstanding Universal Value and integrity. It appeared that the state party wished to pursue hydrocarbon exploration activities within the boundary. You can't win them all.

I returned to Oman 6 months later and was shown around the

construction of the access tunnel and the Al Hoota Cave. The department of tourism asked me to manage the cave but I was not too keen on working in the desert so declined. With the help of Prof Ellery Hamilton-Smith from Australia I did recruit a suitable Australian candidate. The cave was open to tourist in 2006 and appears to be a popular destination. Oman is one of the great success stories of the Middle East with a tolerant population and a great future for tourism.

   I considered it important to understand the culture as the men dress all in white in their 'dish dash a' and the ladies in veiled black. Alcohol was of course originally banned as Oman is an Islamic state but the story goes that aftershave was in high demand due to its alcohol content so the ban was lifted. The bars are full but with no drunks, ladies dance on the stage and the clientele buy them a flowered necklace, a strange custom.

I had always been interested to investigate thermal caves beneath a city which I knew little about, where better than Budapest in Hungary. I suppose I liked the word Budapest as it had Buda in it. Besides hot springs, wonderful people, good beer and magnificent buildings it also had Prof Szabolcs Leel-Ossy from Eotvos University, Budapest. Regarded as the expert on these hydrothermal caves I managed to contact him and arrange a meeting. He had visited Mulu some years previously so seemed to know who I was. I was accompanied by a drinking pal called Joe O'Hanlon, being of Irish descent Joe could look after himself as he spent his time locking up criminals after hauling them off to court. He was not a caver so I arranged some trips with the Professor, his son and friends.

   The wonderful thing about these caves is that they are all totally protected under national parks legislation, the UK seemed well behind. Ferenchegy Cave was discovered during road works which were halted until the Professor and friends could map and explore. It's protected by multiple gates and contains some upward developed thermal tubes and barite crystals. Jozsefhegy was even more fascinating as excavations for a 25-house building site uncovered a hole. Szabolcs spent 85 days of continuous digging to reach a depth of 65 metres to a large passage encrusted with gypsum and aragonite crystals. The building plan was stopped and the complete field scheduled as protected. The explorers even have their own building on the site with changing rooms, store and kitchen. The Hungarians took their caves seriously as they are regarded as national treasures.

25 years was fast approaching since our last trip to New Britain. As usual, being retired, I was becoming bored again and gave another expedition to this amazing remote island of New Britain some careful consideration. Was it possible to go again? I had kept in touch with Jean Paul Sounier who had become legendary with his regular expeditions to New Britain. His reports on logging, palm oil plantations and possible quarrying activities I found very disturbing as this primary jungle karst area contained the largest river caves on earth and was truly spectacular. Could it be saved as a conservation area and bring long term sustainable benefits for the local communities? I believed the answer to that was yes with a great deal of work and initiative it could be achieved. The karst area certainly deserved World Heritage status.

In 2001 at the World Heritage Conference held at the Gunung Mulu National Park, Sarawak, it was suggested that the karst areas of Papua New Guinea could be scheduled as World Heritage sites. Unfortunately this was put on the back burner and no action was taken by the caving community or anyone else. I therefore decided to take the matter into my own hands and began lengthy correspondence with Prof Elery Hamilton-Smith from Australia, South East Asia's foremost karst conservationist. Jean-Paul Sounier and Prof Philippe Audra from France joined the working group which was soon to be added to by Dr Leo Salas, a Venezuelan working for The Wildlife Conservation Society and based in Papua New Guinea. We all agreed that certain areas of karst in Papua New Guinea would qualify for World Heritage nomination. Along with Jean-Paul I drew a proposed boundary for the Nakanai which came to over 3000 square kilometres in extent.

The problem was the Papua New Guinea legislation which stipulated a 'Conservation Area' under the department of Conservation and Environment. In PNG the land comes under the stewardship of the community and is not owned by the state but by the people. Efforts to change this had resulted in riots. The argument was that development was curtailed due to land ownership. This meant that all head men of every village close to or within the proposed boundary would need to agree to the karst becoming a Conservation Area, a massive task. The first thing to do was to get the karst on the World Heritage Tentative List.

A proposal document 'Conserving the Sublime Karst of Papua New Guinea' we eventually published under the auspices of the Task Force on Cave and Karst Protection, World Commission on Protected Areas, International Union for Conservation of Nature. This proposal was a serial site including Nakanai in New Britain and Muller and Telefomin Hindenberg karst areas on the main land.

The document spells out our long term objective of protection and World Heritage status for these karst regions. It was considered that if we do not make an attempt to save the PNG karst, no one else would and the caving community will kick itself for ever more for not taking positive action before it became too late. We therefore succeeded after a few years in having the Nakanai included on the Tentative List. The next problem was how to succeed in obtaining UNESCO World Heritage nomination. The documentation for nomination is a huge complicated task which normally takes a few years with a highly specialised team and integrity needs to be justified. The area requires complete protection under the law and I estimated two years of work. It was hoped that the Department of Conservation and Environment would complete the task but it became evident that this was not to be.

How to obtain funding for a two- year project to conserve the forest and caves of East New Britain? WWF were snowed under with projects in Papua New Guinea, other conservation NGO's also drew a blank. The Darwin Initiative seemed the best possibility as this is a Department of Environment, Food and Rural Affairs (DEFRA) government grant for conservation of £5 million. Application was very complex and it became obvious that a grant would not be awarded to an individual so Graham Proudlove suggested working through the University of Manchester with the Faculty of Life Sciences. I attended an interview and the project was liked as it seemed that I could be awarded a Research Fellowship. We tried the application for two years but were rejected both years, the third year the Faculty decided I was wasting my time. In desperation I even obtained an interview with our Member of Parliament who asked a question in the House, all to no avail. I reluctantly gave up the attempt but at least I had tried. I never did get my Research Fellowship.

18

Untamed Rivers of New Britain 2006

'It is only when we truly know and understand that we have a limited time on earth and that we have no way of knowing when our time is up that we will begin to live each day to the fullest, as if it were the only one we had.'

Elizabeth Kubler-Ross

I began to think seriously about a return trip to New Britain after 25 years especially considering our attempts to obtain protected area status for the Nakanai karst mountain and the ridiculously wet caves. The main problem was how to raise the money? I enquired from a friend from National Geographic Society, Washington DC called Stephen Alvarez if the NGS had ever completed a story about New Britain? The answer was no. I had worked with Stephen during the Gunung Buda Caves project in Sarawak so knew of his expertise in photography and caving expeditions. He suggested I seek an NGS grant and he and some assistants could come as NGS photographers. As this sounded like a good idea I did as it could be done on line. Amazingly I was successful. All I needed now was a definite objective, a team and equipment. This was going to be a long and difficult job but, I had done it before so was confident I could do it again.

The first thing to do was to set an objective. Considering the sheer size of the Nakanai karst area at over 4,000 square kilometres, there were certainly plenty of unexplored locations which all had potential for giant river caves. The southwest zone close to the rivers of Melkoi, Naka, Matuba and Torlu needed thorough exploration. The Bairaman sector, including possible downstream resurgences was also a likely area of interest. The southeast zone with the Rak, Bergberg, and Ikoi rivers looked promising as did the northwest zone including the Ora Cave and dolines to the west. The central north zone close to the Manang River seemed to have a large collapsed doline noted on aerial photographs. The Toiru Basin really needed a reconnaissance to ascertain its potential.

There was no way I would go back to New Britain without the King of the Nakanai, Jean-Paul Sounier from France. Jean-Paul had led more expeditions to the Nakanai Mountains than anyone else and had wide ranging experience of the potential areas for major discoveries. He was first on the list for an invite. I had met him during our 1984/85 expedition to New Britain and also at Gunung Mulu Sarawak many years later. We discussed the possibilities and finally decided on the Ora Cave as the main objective. The twin doline is so large it can be seen on satellite photographs estimated at 900 metres in length and 550 metres wide, situated at the top of the plateau at an altitude of 1,029 metres and high above the village of Ora which is at 620 metres altitude. The village is situated close to the head of the Iso River, a long way from anywhere. Three teams had descended the doline by free climbing 275 metres down a steep slope on the northern side. The Australians in 1973, 33 years ago, the French expedition in 1980 and also my Untamed River Expedition team in 1984. No one had been back since. All three teams had tried to go beyond the white-water rapids about 180 metres in, where both walls were sheer with no banks on either side to be seen. The underground river was around 6 to 8 cubic metres of water flow per second, a formidable obstacle. Before thinking about the very difficult logistics of getting there, I needed a team.

Having spent 10 years "playing" in the Borneo jungle I was a little out of touch with the British caving scene. Who was mad enough to come? The NGS application required three referees so I asked Tony Waltham and Dick Willis. I needed a caver from the USA to back up the application and asked Prof Herbert Laeger, whom I had caved with in Gunung Buda Sarawak and I considered him as an honorary Brit. He asked to come along and I readily agreed. I of course asked Tim Allen and Alan Gamble but both wisely nay respectfully declined. Dave Clucas and his partner from Flowguard Ltd had been of great help during the organising of the Untamed River Expedition in 1984. I had used their photo-copying machine to print out a thousand bits of information. Dave liked the name of the world's largest river cave we explored so much he named his daughter Nare. Along with Dave Sims they had hiked up to gaze at the Nare doline many years ago so he knew what he was letting himself in for. Andy Eavis had asked that if I ever went back to New Britain, could he come. I of course said yes. I now had a team of five old men plus three from the NGS of Stephen Alvarez, Matt Oliphant and

Nancy Pistole. I knew Stephen from our work in Gunung Buda, Sarawak and had met Nancy in Gunung Mulu some years previously.

New Britain, Papua New Guinea. Bougainville shown on this map has now voted for independence

What I needed now were some young crazy tigers. The problem with the NGS is that all photographs are restricted so I needed an independent photographer. The obvious choice was Robbie Shone, young, hard, fit and probably crazy enough to come. Robbie lived not far away in Litton Derbyshire but I did not have his email number so I contacted David Nixon better known as Moose. David was the Derbyshire Tiger with many hairy explorations in the Peak District Speedwell and Peak Cavern systems. He had kindly led me up his climbs in Peak Cavern years ago which I found to be incredible feats of exploration. Of course David asked me why I wanted to know Robbie's email address so I told him. The answer was can I come? which was yes. As far as I knew, David had not caved outside of Derbyshire and had not been on an expedition. I enquired if he could be invited on the next Mulu Caves Project expedition, great practice for slogging through a jungle. He went and loved it with no problems of the heat and all the other nasty things associated with giant caves and jungles. I now had a team of 10 but really needed a few more. Tom Chapman came highly recommended, a high-rise worker with a vast experience of expedition and rope work with an easy-going character. James Alker was recommended as he had been with Robbie in the Gouffre Berger cave in France. The number 12 seemed about right. I now had a team but as usual the logistics became a massive task.

We managed a few sponsorship deals with Lyon Equipment supplying our discounted caving gear plus six Petzl MYO XP caving lights. For mapping two Leica Geosystems DISTO classic 5a measuring instruments were supplied to us along with Suunto three wrist compasses, altimeters and Tandem compass, clinometers. Palm Equipment gave us two throw-bags and cow's tails plus discounts on personnel flotation devices and Helly Hansen discounted LIFA base layer clothing.

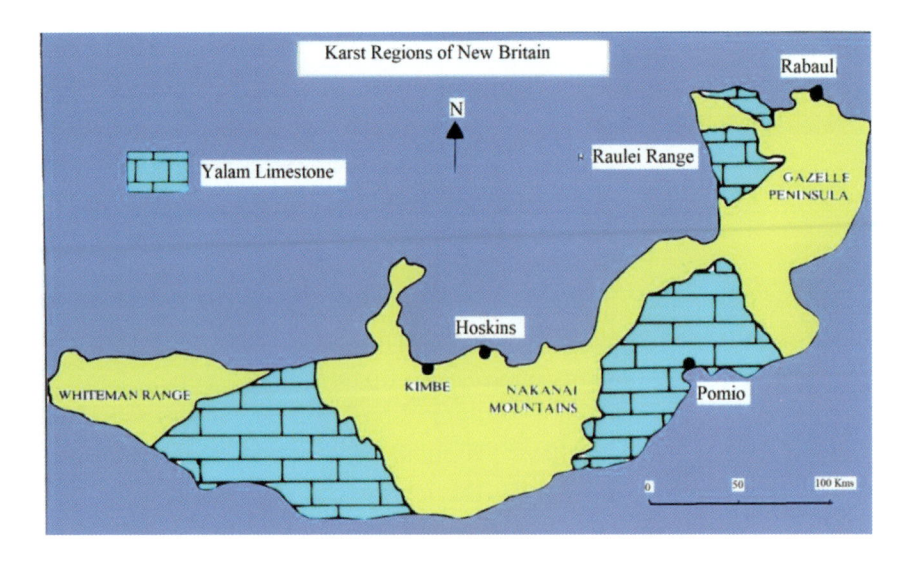

The major karst areas of New Britain, Papua New Guinea

We also received financial support from the Royal Geographic Society and the Ghar Parau Foundation so I finished up with £44,400 which meant we could hire a helicopter. This would solve the logistical problem of access onto the plateau at the other end of nowhere.

I delegated the work load with Robbie concentrating on photography and equipment, Dave Nixon caving equipment, Tom on food supplies and camp equipment and James on medical supplies; which all worked well.

Most people have never even heard of New Britain and have little idea even where it is. New Britain is a part of Papua New Guinea and lies 88 kilometres off the east coast, with an area 36,520 kilometres. The length east to west is 477 kilometres with a maximum width of approximately 80 kilometres. The highest elevation is 2,438 metres and the northern coast is dominated by active volcanoes. The volcanoes are a product of a subduction zone over 9 kilometres in depth along the south coast, frightening to swim in crystal clear water over a 9 kilometre deep cliff edge.

The island has therefore been undergoing vigorous uplift since the Late Pliocene and experiences regular volcanic eruptions and earthquakes. It is common for every expedition to the Nakanai to experience violent earthquakes. Between July and September 1994 Rabaul, the capital of East New Britain, was almost entirely covered by volcanic ash from the eruption of two neighbouring volcanoes, for the third time in its brief existence of less than 100 years. As much of the capital and airport was destroyed the administration centre was moved further south to Kokopo and a new airport constructed. This was sad as I liked Rabaul with its once lush gardens and white bungalows. Such violent seismic activity and regular explosive eruptions typify the present tectonic setting and it is regarded as being the most highly seismic area in the world, thankfully with little effect on the caves. To a high degree the speed and extent of New Britain's cave development reflect the island's extraordinary tectonic regime.

Seventy separate languages are spoken among the population of around 514,000; most live in the coastal regions engaged in subsistence farming and fishing, or working on plantations. The local population around the village of Pomio are from the Mengen tribe while, up in the mountains, they are Kol tribe's people. Animosity between the tribes had now ceased so it was reasonably peaceful.

Many small villages situated in the hinterland exist on subsistence farming and rarely come into contact with Europeans. Sweet potatoes, taro, mangoes, sugar cane and strange green vegetables are the staple crops. There is little meat in their diet except for the very occasional pig, so malnutrition can be a problem during a poor harvest. In an expedition of this nature I considered it essential to have a basic understanding of the culture of the local community due to problems with land rights. Having been closely involved with the community for five months during the 1984/85 Untamed River Expedition I had a basic understanding of their beliefs, traditions, black magic, the cargo cult and the problems faced by the day-to-day living conditions. As for Betel nut chewing-known as Pinang in Papua New Guinea which comes from the palm tree *Areca catechu*-I left this alone as, taken with powdered lime they spit the red juice everywhere. It's a mild stimulant akin to caffeine or tobacco. Walls and streets are covered in blood red blobs with signs in administration buildings saying 'no chewing'. I had slept in their huts, eaten their food and walked for many kilometres with them through the jungle. It was essential to be trusted as they could, if desired, make life very difficult for foreign strangers. During the last expedition I had handed out jumble sale clothes as a sign of friendship which worked out well. Long newspaper rolled cigarettes were very popular as gifts. It was going to be interesting to see if anything had changed over 25 years, surprisingly little had.

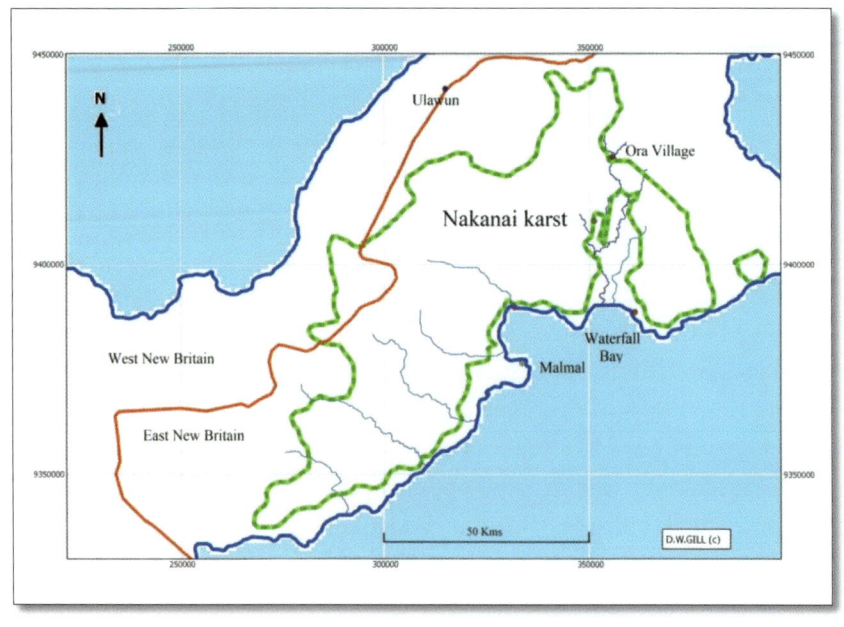

The Nakanai Karst Area, East and West New Britain

The island has two provincial areas, West New Britain and East New Britain. Most of the karst of the Nakanai Mountains lies in East New Britain, with a small section within West New Britain. East New Britain has an area of 15,724 kilometres, with around 4,300 kilometres of karst. The island is tropical, with a temperature of 23–31°C and an extremely high annual rainfall of 10 metres to 12 metres. This is the reason for very wet caves containing momentous rivers and the extreme up-lift rate of the mountains due to the tectonic movement of the plates. Many of the underground rivers are at high altitude having been uplifted faster than the down-cutting. Resurgences, where the underground rivers emerge into daylight, tend to be high up in the steep-sided cliffs of the surface gorges. Mountain areas are subject to an oceanic monsoon climate and the significant and continuous average high humidity of 80%, combined with warm temperatures all year round, has enabled development of rainforest that is still largely untouched by industrial exploitation. Such rainforest has a significantly high biodiversity and an environmental study in 2009 identified more than 100 animal species previously unknown to science in the Nakanai mountain range. Logging and palm oil plantations, widespread around the coastal regions, are perceived as the major threat to the integrity of the island.

The Ora doline is situated in a very remote part of the mountain range and it would most probably take over two weeks before the team could establish a base camp there. Over 150 square kilometres of unexplored sinks, dolines, blind valleys and resurgences awaited exploration in this area which, as far as we knew, had not been explored by anyone as the local population rarely travels more than a few kilometres from the village.

Deciding on the logistics was not as difficult as expected as working on the World Heritage Tentative List I had made a number of important contacts. Florence Paisparea was the East New Britain Provincial Government Environmental Officer and proved to be of tremendous help, along with Maureen Ewai from Conservation International. I also made contact with Peter Lavender the base manager of Niugini Helicopters and James Robins Divisional Head of the National Research Institute. We now had official permission and finding out as much information as possible on Ora village, I had a lucky break as there were Europeans living there.

It appeared that two families of Australian and USA missionaries had a base at Ora village. They were from the New Tribes Mission teaching Christianity, a strange group that established bases at remote locations. These bases seemed to have been established without government permission and there were a number of conflicting reports as regards to their operations. Their head office was in Hoskins on the north coast said to be surrounded with barbed wire. It also appeared there was no shortage of money as they built their nice houses at remote locations with helicopter delivery of materials and food supplies and they had their children with them. Every year they went on a sabbatical to their home country giving talks and raising money. They had radio contact via their head office so I made contact to give them the details of the expedition. Their answer was for us not to come. As we had permission by now from the Provincial Government I will leave my answer to the imagination.

I now had the team, objectives, budget and logistical plan; all that was needed now was to go! The dates were set from 11th January 2006 to around the end of March giving us over two and a half months in the field.

The secondary long-term objective was to work towards the establishment of the Nakanai Mountains Conservation Area, eventually to propose the area for World Heritage status. The expedition's involvement would consist of meetings with the relevant government departments at provincial and local level; NGOs; local village head men; villagers; logging company management, and the collecting of relevant information on the state of conservation. Prof Elery Hamilton Smith from Australia would join us in Kokopo for negotiations with the Provincial Government officials.

The journey out was uneventful and we met up with our National Geographical Society USA team in Singapore with Matt negotiating a deal for our equipment access baggage claiming it to be diving gear.

While the majority of the team were engaged in meeting relevant government officials in the capital of Papua New Guinea -Port Moresby and later in Kokopo to purchase supplies, I sent off a two-man reconnaissance party consisting of Andy Eavis and James Alker. Their mission was to establish the best route in. Sadly Andy could only come for one week which seemed a terrible waste of time, energy and money considering it would take us over two weeks to establish a base camp on the plateau. They were to make contact with the logging camp at Matong shore base and with the two Australian and American missionary families of the New Tribes Mission, stationed at Ora village. Ora village is situated at the headwaters of the Iso River, in one of the remotest areas of the Nakanai Mountains. The old logging road from Pomio village on the coast of Jacquinot Bay reaches all the way to Nutuve village, but this was reported as being overgrown and with many bridges now in a state of collapse. The original plan was to walk in to Ora village from Nutuve village or Nutuve logging camp, on a trail around 22 kilometres long. After flying from Kokopo to Palmalmal situated close to Pomio and crossing Jacquinot and Waterfall Bay by boat to Matong logging camp, a Niugini Lumber logging truck transported the reconnaissance team 35 kilometres along the well-maintained logging road which leads north to the Nutuve logging camp. This camp is situated between the Berg Berg and Ikoi Rivers, 6 kilometres east of Nutuve village. The road lies mainly east of the Berg Berg River and crosses it 30 kilometres upstream. From the logging camp, a two-day walk via Nutuve village took the team to Ora village where contact was made with the missionary families. Communication with the main team in Port Moresby was established via satellite phone. The walk was described as hard and required the crossing of two major rivers; the Ikoi and Essis, both of which ran through deep steep-sided gorges. Although there were many villages along the way, porters were few and far between. From the reconnaissance report it was obvious that the only option was to use a helicopter to transport the main team; along with its three tonnes of equipment and food supplies, from the Nutuve logging camp to Ora village.

We met Elery and meetings took place in Kokopo to enable real progress within the plans for conservation with the Department of Environment and Conservation; the National Research Institute; the Provincial Government of East New Britain; the Governor of East New Britain; the Tourism Authority; later the Local Level Government Pomio District; non-government conservation organisations; as well as with local village headmen; villagers, and logging company management.

Two team members Dave Nixon and Robbie Shone flew by helicopter direct to Ora village from Rabaul, one member of the reconnaissance party Andy Eavis returned on the empty helicopter as he had to return to Britain. Sadly he did not manage to see the giant Ora doline before his trip back. The three team members remaining at Ora then climbed up the 400 metres cliff to the plateau and located the Ora doline after a three-hour walk across numerous blind valleys and dolines. Over the next few days a base camp was built and a helicopter landing zone with the help of Ora villagers, it overlooked the Ora doline. I considered it essential to have a contingency plan on the plateau in case of emergency evacuation.

The Ora doline and base camp. Photo: David Gill

A freighter was hired from Rabaul to ship the team and our supplies around the south coast to the Matong shore base logging camp, a two-day journey. From the logging camp trucks transported the team and equipment to the Nutuve logging camp. Accommodation was kindly supplied by Niugini Lumber, a Sarawak-based company. The managers were from Sibu, Sarawak so, considering my close association with the Forest Department Sarawak we had no problems with accommodation in their huts and food. They refused to accept payment for their services. One week later after a delay due to heavy rain, we arrived at Ora village by helicopter from the Nutuve logging camp with all our supplies. On the way our pilot flew us along the Iso gorge where we were amazed to see a massive multiple waterfall emerging from caves over 100 metres above the raging white-water river below. The locals called this Mageni, an obvious secondary choice for exploration. Considering this was 2006 it seemed strange that this wonderful magnificent falls had never been recorded or photographed previously; it really was unexplored country.

The Mageni Resurgences overlooking the Iso Gorge. Photo: David Gill

Sadly the helicopter could not land safely at base camp on the plateau as the pilot claimed the logs laid out were the wrong way around and he was afraid of getting his skids trapped between the logs, we were not qualified to argue the case. He was a cautious pilot as some years previously he had sustained severe injuries in a crash.

Ora village. Photo: David Gill

The two missionary families based at Ora village were very friendly and accommodating; they administered to a population of approximately 100 Kol people and they kindly provided storage space, accommodation in the village and translated from Kol to English. This greatly facilitated the hiring of porters which included women and teenagers. Over a four day period all the team, together with our equipment and supplies; finally established themselves at the Ora doline base-camp; it was 27th January, 17 days after leaving home. 160 porter-loads had been required carried from the village. For the majority of the time on the plateau two men from Ora village on rotation stayed with the team at the base camp which was normal procedure considering that the area was their customary land. They were of course amply fed and paid. A good working relationship with the local community had thankfully been established without too much trouble. At last we could start exploring.

Dave Nixon and the tigers wasted no time and began searching through the highly vegetated karst to the north-west with numerous sinks and blind valleys. They headed in that direction for obvious reasons, the projected route of the upstream underground river. We knew that the Ora River was flooded upstream after a short distance. Any draughting sink could lead into the upstream continuation of Ora.

Ora base camp. Photo: Robbie Shone

Many sinks were investigated, but most proved to be immature; they are either too narrow, or are blocked with sediment. It appears that there are so many blind valleys; dolines and sinks on this area of youthful plateau that drainage is distributed into numerous small catchments, with little scope for large surface streams to develop and drain into larger sinks. Some of the gullies contain mudstones over a metre in depth, which may be derived from old volcanic eruptions. This area is closer to the volcanoes on the north coast of the island, so may account for the large amounts of decomposed ash. The Nare Cave region far to the south has far less soil cover. Many of the caves in this region are fed via well developed canyons taking large wet weather streams, thus have large cave entrances. The extremely rough karst terrain makes traversing it very hard work. Tree cover on the ridges between the dolines and blind valleys has been severely damaged by Cyclone Justine in 1992. Of the larger trees, 60-80% has been destroyed. The ridges therefore tend to be overgrown with new tree growth, bamboo and fallen trees. Cutting tracks on level ground is rarely possible, as the only way to progress is to descend into each steep-sided enclosed valley and climb up the other side, immediately to descend into another one.

Challenger Pot, at an altitude around 1000 metres, was explored over the course of the expedition for 143 metres in length and 72 metres in depth. Pulse Pot was another at 60 metres in length and 46 metres deep. These were just two of the many immature sinks feeding the upstream Ora Cave System.

Triosaurus Pot was found to the south with a hoped-for route into downstream Ora. The depth was 62 metres but the cave was blocked with clay and debris.

Hundreds of square kilometres of karst in this area remain totally unexplored by man, as even the Ora village people do not venture onto the plateau. Vegetation within the valleys which escaped the cyclone is in pristine condition; rich in epiphytes; palms, and gymnosperms. Various teams spent six days cutting a track to a large blind valley to the south of Ora that had been spotted on the aerial photographs, but they only covered 2 kilometres as the crow flies. It is a rare privilege in today's world to be able to traverse large areas completely untouched by man. At times it was akin to being in a lost world expecting to meet a dinosaur in the next blind valley. The major find was Phantom Pot which seemed to be a good possibility with a strong draught.

An easy route was found to descend the Ora doline without the need for fixed ropes on the north-east side but the Ora River looked to be in high flow, approximately 8 to 10 cubic metres a second, far higher than previously reported.

Base of the Ora doline. Photo: Robbie Shone

The left hand wall of the downstream entrance portal was sheer with fast flowing white-water, but a bank could be seen on the far side of the river. The upstream entrance was followed via a short cascade to a point where the river could be crossed via a steep-sided canyon by a very long-legged man, in this case Matt Oliphant. The right-hand bank was then followed downstream to gain a fixed anchor point for the tyrolean rope at the downstream entrance. A low level tyrolean was installed to cross the river above the falls, later to be replaced by a much safer high level tyrolean well above water level.

The right-hand bank soon gave out at a sheer wall and a further very wet tyrolean crossing was established in the river to reach a large washed-in log on the far side. This second tyrolean was achieved using the trail ferry technique of hooking a grapnel behind the log and being swept across by the force of the river. Techniques used to good effect in the Nare Cave River in 1984.

A short section of bank followed to more white-water. Both sides of the passage appeared sheer with fast flowing water disappearing into the distance. This was presumed to be the furthest point reached by the Untamed River Expedition in 1984 when the river must have been in fairly low flow. A ledge high up on the right looked a possible way on so the river was crossed once more and a climb reached the ledge. Tom climbed to the ledge where a peg was found; presumably left by the 1972-73 Australian expedition over 30 years previously but still good enough to use for aid. It is possible that the Australians failed to reach the ledge otherwise they would have probably explored the cave to a conclusion.

Tom Chapman crossing the 1st tyrolean. Photo: Robbie Shone

At roof level, Snake Ledge was rigged with traverse ropes gradually descended back to river level. A further short roped traverse along the sheer wall terminated at some nice gour formations. The way on was again at river level with fixed traverse ropes. A fourth tyrolean followed to reach the left-hand bank. A short section of wading unfortunately terminated in a sump after a total distance downstream of 630 metres.

Little Ora River Cave issued as an inlet from a low passage at the downstream entrance. This immature streamway was mapped for 273 metres to a 14 metre climb with the cave seen continuing beyond.

The upstream cave was explored for a total of 315 metres into the south doline. A small overflow stream entered from a cave entrance on the west side of the doline and was followed to the main river and a large sump infested by small flies. The main river disappeared into a siphon to the north east of the chamber reappearing in the main upstream passage.

The 3rd tyrolean up to Snake Ledge. Photo: Robbie Shone

The total length of the Ora River Cave system came to a distance of 1,220 metres, with a depth of 317 metres from the lowest point on the doline rim.

Ora's twin dolines were surveyed and proved to be a staggering 1400 metres in length, 750 metres across and 200 metres in depth-placing Ora among the largest collapsed dolines in the Nakanai Mountains and in the world. Throughout most of the four weeks of exploration, the Ora River rarely reduced in size from 8 to 10 cubic metres a second of white-water rapids.

As usual with a British caving group of this nature in such a remote corner of the earth, humour was prevalent. The earthquake was notable when late at night the whole camp shook alarmingly and threatening to collapse. The kitchen fell to pieces and people fell out of bed. It was recorded at 5.8 on the Richter scale. The aftershock was just as bad.

The gour pools beyond Snake Ledge and the 4th tyrolean with the sump just beyond. Photos': Robbie Shone

Rain was as usual prevalent so a great deal of effort was made into collecting water and refurbishing the tarpaulin roof as large areas produced collecting points with massive pools of water threatening to collapse. I woke one night to find a large half naked man straddled across my bed moving in an up and down direction inches away from my face. 'I know you have been a long time in the jungle Tom but this is ridiculous' I said. 'Just trying to get rid of all this water above your bed, it's about to burst through' he replied. I forgave him his little indiscretion.

The funniest episode has to be the story of the cheese. I had ordered a few kilograms of cheese from Tropicana our food supply store in Kokopo but they had run out of stock. The manager promised to pass on the cheese to Peter Lavender from Niugini Helicopters when his stock arrived. Peter promised to pass on the cheese to any helicopter passing our way as they regularly flew from Kokopo to Hoskins on the north coast. One day a passing helicopter spotted our base camp and out of curiosity managed to touch down on our logged landing zone. The pilot was immediately accosted by a grubby looking half-naked unshaven white man brandishing a machete demanding his cheese. As he had no idea what he was talking about the pilot took off rapidly convinced this very remote jungle at the top of the plateau had a camp with a group of strange half-naked crazy English drop-outs demanding cheese. Unfortunately we never did get our cheese.

The insignificant entrance to Phantom Pot was discovered during the early stages of the expedition and gradually explored over the course of our stay at base camp. Although little over 1 kilometre from camp it took almost one and a half hours to reach the entrance, giving some indication of the severity of the karst rainforest on the plateau.

At an altitude of 1045 metres, 1400 metres of nasty, sharp, meander passage, "The Meanders" with a pitch of 18 metre plus a further 6 metre and 20 metre pitch, finally entered the upper Ora River passage, "Upstream Endurance". This large river passage was explored upstream to a massive lake, "Lake Myo" 67 metres long by 58 metres wide, the chamber being approximately 60 metres high with a 10 metre high waterfall "Myo Falls", pouring into the lake. This was climbed by Dave Nixon with the help of the Australian piton found and retrieved from Ora Cave. This less mature river passage, "Acoustic Canal" was explored for 800 metres, eventually leading to another siphon.

Lake Myo with the main stream inlet and Dave Nixon climbing Myo Falls. Photos': Robbie Shone

A large fossil inlet passage leads off from the lake for 400 metres passing a narrow inlet, "Zigzag Alley" 341 metres in length. The main passage terminates at a pit blocked with debris. This had clearly been the active river passage before being blocked by a collapse that diverted the main river to its present course. Fossil passages are unusual for the Nakanai caves, as the majority of cave passages so far discovered still contain active rivers.

Downstream Endurance Passage with the midge flies.
Photo: Robbie Shone

Downstream a fine 40 metre by 40 metre river passage in deep water leads to a siphon after 200 metres. The separation between this downstream siphon and the upstream siphon of Ora Cave is 132 metres. The exploration in "Downstream Endurance" was plagued by millions of small midge-type flies, the same as found in the upstream siphon lake of Ora Cave. Phantom Pot was surveyed for 3.9 kilometres in length, over a depth of 191 metres. This cave can only be described as extremely difficult in nature to explore, far too difficult for a man of beckoning age like me.

Further surface exploration to the north east may well find a continuation of the upper Ora river beyond the present end of Phantom Pot, but the terrain is extreme and logistics considerable. The potential for gaining access into the further reaches of the upstream Ora River cave system is good, but would require a mammoth effort in track-cutting, with many bivouacs. Tens of kilometres of upstream river cave probably remain to be explored.

Disappointingly Dave Clucas and Herb decided to arrange for a helicopter to fly them away, they'd had enough of the rigours of the jungle. This left us with a team of nine and a decision was made to vacate the plateau as there was little more that could be achieved.

The Ora Resurgence and Endurance Passage. Photos': David Gill and Robbie Shone

We would split into three teams, one to look at the Ora resurgence and the second a reconnaissance to the Mageni waterfalls. The third was Tom who agreed to stay with me and dismantle the camp, plus arrange for numerous porters to carry all equipment and spare food back to Ora village. Dave, Robbie and Jean-Paul with a group of porters would descend the mountain and with food supplies, caving equipment and attempt to locate the position of the Mageni waterfalls and set up a bivouac camp. James and the NGS photographers were to camp at the village and attempt to climb up the cliff to the Ora resurgence entrance and explore beyond.

Thanks to Tom's dismantling of the base camp, this worked well as did the arrangement of all our porter loads. While there I received a visit from a group from Ire Village near the Nare doline. They remembered me with affection as they were children in 1985 and had heard I was in the vicinity. They had walked all the way and climbed up the plateau just to see me. They wanted me to visit the Nare once more which was a good idea as Stephen Alvarez wanted to photograph for NGS this amazing doline as it is one of the world's greatest speleological wonders. I gave them our tarpaulins and some money to build us a small base camp overlooking the doline. They departed happy. It later transpired that the camp had been dismantled by an opposing group from another village, claiming the Nare doline was within their traditional land, a problem that did not exist in 1985. The problems with land tenure still existed so we decided regretfully to abandon the idea rather than be responsible for starting a violent conflict.

One family of missionaries had left their house in Ora village to conduct their sabbatical so we had the use of a store and sheltered sleeping on the veranda without having to build a camp.

James completed the 40 metres climb of the Ora resurgence with the NGS team taking some great photographs; he explored a large passage sadly to a siphon after just 129 metres.

The total length of the Ora caves, including a few small and relatively immature caves feeding into the main drain, came to 5.4 kilometres. After all these years the exploration of Ora Cave and associated caves had at last been completed, a highly successful trip thanks to an amazing team.

A three-hour walk south from Ora village located the Mageni falls- which could be clearly heard from the edge of the cliff. An abseil of 50 metres and a traverse led into a cave entrance with a large river issuing from it of approximately 3 to 4 cubic metres a second, a substantial river. Tom joined the group and later James and the NGS team. Returning porters to Ora village passed on messages to me as regards to progress and equipment requirements, I sent them back heavily loaded over a few days. I joined the group and settled into a very basic but adequate bivouac camp right at the edge of the cliff.

Descent of the cliff to Mageni Cave. Photo: Robbie Shone

The team hit a waterfall which was climbed by Dave and led uphill with river passages, oxbows, waterfalls and some highly decorated galleries. The cave was mapped over a period of ten days with some long arduous trips for 7.2 kilometres with a vertical range of + 229 metres. The Mageni cave system still has open river passage at the furthest point reached, a strong draught and a major flow of water. It drains from the northwest, from an area beneath the large valley seen on the aerial photographs. This major river cave system lies parallel to the Ora river cave and there are probably many further kilometres to explore beyond the furthest point reached. This would be a major undertaking probably requiring an underground camp. Like all the underground rivers in this region; including Gamvo and the mighty Nare, they drain from the northwest.

The Mageni steam way, waterfalls and Derbyshire Downfall.
Photos': Robbie Shone

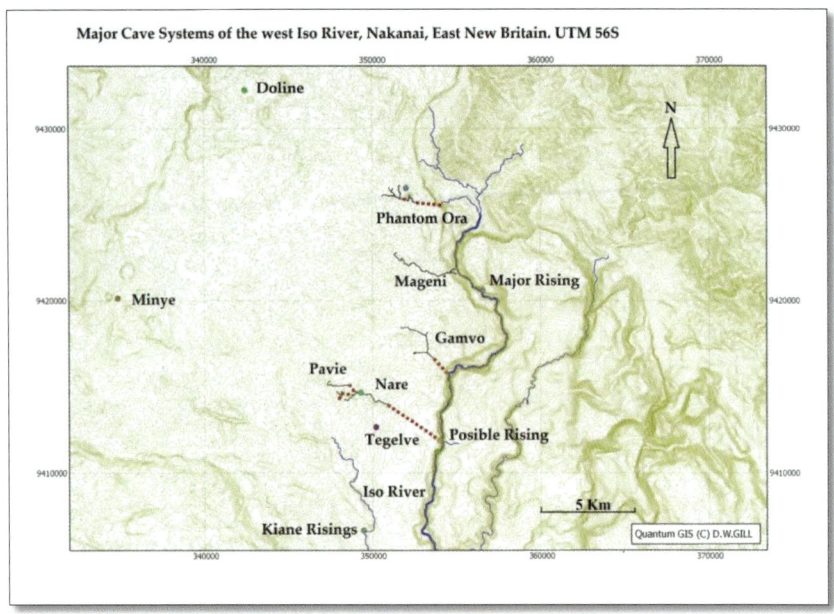

The Ora, Mageni, Gamvo and Nare Cave Systems

I had arranged for the helicopter to pick us up from Ora village so after 10 days we returned to the village only to find out the chopper had broken down. A long wait ensued which was annoying as we could have spent a few more days at Mageni. At last the helicopter arrived and we donated our spare food and supplies to the village and were dropped off at the shore base logging camp. As the Nare trip had been cancelled Stephen required more photographs of a massive cave, Kavakuna was the ideal place so we hired a boat and settled into a guest house in Pomio. Stephen, the NGS team and Robbie hiked up to the entrance and camped for a night obtaining some wonderful photographs.

Another large doline spotted on the aerial photographs lies 20 kilometres west of Ora. This was flown over by helicopter, and a river was seen crossing its floor and flowing into a cave entrance. This area is normally obscured by clouds so the doline had not been recorded previously. This cave was explored by a later French expedition but soon hit a siphon. It appears that Ora may not be the last of the great dolines and river caves that remain unexplored in the Nakanai Mountains.

The National Geographic Society Magazine article was published in September of 2006 and prompted the BBC Natural History Unit to request a film of the Mageni Cave "The Lost Land of the Volcano" in April of 2008. The author, Jean-Paul Sounier and Dave Nixon took part on this expedition as we were getting paid, although not much. A further 2.2 kilometres was added to the total length of the Mageni Cave bringing the total length to 9.4 kilometres.

The expedition explored and mapped a total of 12.6 kilometres of caves, considering the remoteness this can be considered as highly successful. As can be seen from the map, there is a great expanse of blank areas to the west, a major undertaking to explore.

19

BBC filming of Mageni Cave, New Britain 2008

Humanity's behaviour suggests intelligence is an evolutionary dead end.

Wayne M. Schmidt

On returning back to the UK it was not long before I was in hospital with a severe case of Malaria. It seems I was the only one to catch it, possibly being more susceptible after having had Malaria in 1985 during my last visit. Maybe the tablets were not as effective as they should have been. It is estimated that one million cases per year are recorded with 800 deaths in Papua New Guinea with 18% of deaths in hospital due to Malaria. Worldwide this is still a major cause of concern with 219 million cases recorded in 2017 with 435,000 deaths. After catching it for a second time it seemed much more severe than my first encounter. The logging company we stayed with in 2006 had a huge bottle of anti-malaria tablets they give out to the locals. The missionaries at Ora claimed there were many deaths due to this nasty Plasmodium parasite. These probably go unrecorded.

It was not long before I received a call from the Natural History department, BBC Bristol. They wanted to film as part of a documentary series a cave in New Britain. This resulted from instructions from someone in power at the top after reading the National Geographic Society Magazine article and viewing the splendid photographs. They came to see me in Manchester and I explained the difficulties of filming the Nakanai Caves as they contain ferocious underground rivers. I suggested Ora Cave as this would be perfect for filming white-water underground rapids with some great river crossings. No, they wanted to film Mageni, not ideal but they insisted. They probably liked the photograph of the multiple waterfalls issuing from the cave entrances.

I was invited to Bristol for a meeting and took my wife Betty along but on leaving the BBC a lady, who appeared to be in charge of finances, showed displeasure-explaining that it was not necessary for my wife to accompany me to meetings as the BBC had to pay for our train fare. Considering that I had attended the meeting free of charge I began to have serious doubts as the budget was reported as being a million pounds and they were complaining about a train fare.

I was never certain what my job was supposed to be but I explained that it was essential that the original exploration team came along and I would manage the base camp and liaise with the local population. The Kol tribe can be difficult to deal with but I was known to them and trusted with a good working relationship.

I worked out a feasible itinerary and supplied them with names and addresses of the officials that we needed to consult; also an accomplished underground camera man was required but again they did not seem to take the advice seriously. Keith Partridge was to be their camera man who was a highly qualified individual with a great track record. Keith had been the camera man for the documentary film 'Touching the Void' which won a BAFTA award. Also the BBC decided on a presenter who would come along, he would be the usual super hero facing the dangers and challenges, flexing his muscles and expounding his opinion how amazing and difficult everything was. Pam and Tim Fogg would be their work-at-height safety experts. Considering I had caved with them on previous expeditions, this seemed like a good idea as they are highly competent.

Over the next few months around 200 emails later and other meetings in Bristol, I arranged for the helicopter landing area to be cleared by the Ora village Kol community. This was situated on level ground 20 minutes away from my recommended base camp which was just above the cliff. All accomplished via my contact with the missionaries based in Ora village. They had radio contact with their base camp at Hopkins enabling messages to be passed on.

The team was chosen by the BBC and Tom Chapman-much to my disappointment-was not invited. After a night-time white-water training meet in Wales Robbie Shone pulled out; highly disillusioned. Dave Nixon was still keen as was Jean-Paul Sounier. Strangely the BBC asked Jean-Paul to recruit one of his French friends called William to replace Robbie. Again this seemed peculiar considering the large numbers of highly experienced cave explorers in Britain all capable of managing the rigors of the Nakanai Mountains. The catchphrase became, 'William, who the hell is William'. Again recommendations were ignored as I tried to arrange a practice filming in a wet cave in Mendip. 'No', they said, 'we are filming in a swimming pool'. When I pointed out that caves were dark, they said they would film with the lights out! It began to appear that this trip was going to be a shambles and I was dealing with some very strange individuals.

A free trip to New Britain and a film would, I hoped, greatly expand the possibilities of obtaining total protection for this world significant region and at least I was getting paid a little. I therefore considered it well worth the effort. I was wrong on most points.

I viewed some of their past films in this series with immense disappointment as they tended to film a man and a woman filming! One shot was of a woman filming at the top of a tree on a specially built platform where she filmed a Hornbill bird, hardly groundbreaking as I sit on my veranda of our house on the Melinau River, Mulu and watch the Hornbills squawking in the trees. The one on Gunung Mulu was appallingly bad with a descent of Solo shaft which was claimed as a previously unexplored chamber. When I was working in Mulu a friend and one of my guides had descended the shaft to a conclusion. I wondered why the guides did not tell the film crew, but there again this was the BBC so they probably would not listen.

They offered me the same pay as the other non-BBC members of the team. Considering I had put in over three months of work I asked for a rise and managed to obtain a little more which can hardly be described as generous.

I met up with Dave Nixon in Piccadilly railway station, Manchester on Friday, 28th March 2008. Dave was lugging around the cave rescue stretcher besides his own personal equipment. We met the rest of the crew at Heathrow airport and caught a flight to Singapore, then to Brisbane and finally Port Moresby the capital of Papua New Guinea. Why the BBC decided on this tortuous route, I have no idea as it took three days. As normal with the time zone and not being able to sleep on a plane, I did not get much sleep that night so spent some time helping Keith with his filming and sound recording equipment. At last on 31st March we flew to Palmalmal on a chartered flight with a mountain of equipment and spent the night at the Koki Guest House. I made contact with Alois Magogo who lived in Palmalmal and acted on behalf of the Pomio District Administration. He kindly made attempts to obtain some ladies who could accompany us and act as our cooks but unsurprisingly no one fancied the job so Alois offered his services; this was accepted by Jonny Young who appeared to be in charge. I was also in touch with Peter Lavender the base camp manager for Niugini Helicopters and he arranged our helicopter transport to our landing zone and had also arranged for our required four drums of helicopter fuel.

The following day we set off with the usual mountain of filming equipment, camping, caving and food supplies. The Ora village community had cleared the landing zone just in the right place as I had requested. Jean-Paul and William decided that they wanted to build our base camp next to the landing zone but I had planned to have it built 20 minutes down the steep sloping bank of the valley near the edge of the cliff-just 5 minutes away from the descent point to Mageni. This was a nice level area but required porters to carry our equipment down the slope. As seven Ora people had arrived this was not a problem. Tim, Pam and Dave agreed so we won the debate. I jumped back on the helicopter with the film crew and we were dropped off at Ora village.

Diplomatically it was essential that I clarified with the local community an agreement on employment and pay. I received a nice friendly welcome when I arrived and that night we slept on the veranda of Craig's house. Craig was one of the missionary families based at Ora village. The crew filmed a meeting with the presenter and the Ora head man, fictitiously arranging permission to descend and film the Mageni Cave which I already had arranged. A fee was paid to the headman for their work on cutting the landing zone and porters were also paid 10 kina per day. Everyone seemed happy except for one of the chiefs who handed me a letter demanding 5,000 kina as he claimed Mageni was his cave. Of course we did not pay it as this would form a precedent for future expeditions.

The following day was my daughter Racheal's birthday so I managed on the satellite phone to wish her a very happy birthday. I marvelled at the wonders of modern science and technology. We set off on the two-hour trek to the landing zone with twelve porters in the rain, filming as they progressed. Tim and Pam camped out the night at the landing zone to guard the equipment but next morning began recruiting porters to carry all the loads to our planned base camp. In heavy rain the track down the hill was a streamway and the camp was chaotic with around 50 local Ora villagers hanging around attempting to gain some shelter from the downpour.

The next few days were busy building a shelter for our two men from Ora village who would stay with us, building a work-table, a toilet, generator, wiring and a few lights around camp. Keith required a box building with a light inside in an attempt to dry out his very damp and highly expensive camera.

Jean-Paul and William rigged the rope descent down the cliff to the cave entrance while Dave helped Keith with some filming. Alois and I did the cooking on a primus stove on the floor in the mud.

Unloading the mountain of equipment in the rain, Mageni landing zone. Photo: David Gill.

One hour's walk to the south were a few huts where the occupants, along with Ora villagers, would walk over to our camp with a few fresh local vegetables for sale. Cau cau, corn, pip pip, ferns, spring onions, aibika and taro root all became part of our diet. This helped to break the monotony of tinned food.

That night Dave had a fever and sat by the cliff at 3a.m. but thankfully recovered within a few days. Pam was also sick with a fever for a couple of days but again soon recuperated. Tim worked hard to complete a hauling system in order for equipment to be lowered down the cliff so some filming was carried out in the cave as far as the first waterfall. Problems of course occurred as suspected due to the constant high humidity and a very wet cave. The sound system malfunctioned so I lowered a few more of the transmit-receivers down the cliff. Over the next few days they filmed as far as Derbyshire Downfall and the film looked good. A downstream passage was discovered and filmed being around 200 meters in length. The usual jungle injuries were prevalent with Mulu foot, that rotting of the feet due to being constantly wet and me with a cut hand from the razor-sharp limestone.

By the ninth day of the trip a further 500 meters of unexplored loop passage was mapped but the film of Derbyshire Downfall was badly fogged so it would need to be filmed again with a dry camera. I rigged up the Mole phone where it was hoped we could communicate through the limestone to the cave passage beneath our feet but had technical difficulties with it.

A decision was made to camp underground so Jean-Paul, William and Dave checked out the canal passage and beyond to make sure the planned camp site was not underwater as it seemed to rain every day. The problem was that this was April and the monsoon starts around May when it rains horrendously every day of the week. It would be impossible in May to land a helicopter and a 50 kilometre walk back to the coast with all the equipment in the monsoon conditions was out of the question.

I left the generator running all night in an attempt to dry out the camera in its wooden box-all to no avail so arrangements were made to have a new camera sent out from the UK to New Britain at great expense. This was due to arrive on the 15th April. Filming with a water-proof housing was tried and had some success while I was busy soldering bits of broken wire for the sound systems and filming communication via the Mole phone. The following day some good film of the Derbyshire Downfall looked very spectacular but the Mole phone refused to work adequately with very poor communication.

By the 14th April Keith was in a bad way with a very swollen and painful hand. Advice via the satellite phone was to send him back home for hospital treatment. The original plan was for the filming team and

Dave to camp about two hours upstream beyond the falls while Jean-Paul, William and I would transport the equipment to the underground camp and return to base camp. Unfortunately due to Keith's injury this plan had to be abandoned. Instead I was filmed with Alois explaining the long term goal of World Heritage nomination and conservation of the Nakanai Mountains and its remarkable caves. Sadly this was omitted from the final two part TV documentary. As this was the main reason I was there it came as a disappointment, but there again this was the BBC.

The following day the helicopter arrived at our landing zone and Keith was airlifted to Kimbe on the north coast and from there to Port Moresby which was bad news for the underground filming.

We managed to persuade the Ora children to collect insects for 2 kina per child and bring them to camp where we could film them; this proved to be highly entertaining. The film crew and Dave set off for the underground camp while Jean-Paul and William helped with transportation of the equipment and explored and mapped yet another crystal-lined fantastic oxbow for 400 meters; returning to base around midnight. The film crew only camped for one night as the presenter claimed sickness as from the viewing of the film it became obvious he could not sleep or settle in the cave. The canals are low and flood prone so are intimidating but the film looked gripping.

That night I awoke at 3a.m. to a very loud bird calling, Alois explained that this was a jungle warning. There were bad bush men in the area, not long after he was proved correct although they could not be classified as bad bush men. The following day a large group of around 30 arrived including women and children. The head man Albert Butong claimed that they were from Hobu village to the south at the other side of the Iso River. They had walked via Mugo village and claimed they were related to the Ora clan. I had a long and protracted conversation with the head man via Alois who acted as interpreter. Albert had stopped the loggers from entering his area and explained about the large limestone unexplored mountain on the true left bank of the Iso River. He claimed to have seen a large deep doline with a river at the base flowing into a large cave he called Loduku. He claimed the river emerged from a rising at the side of the mountain and flowed into the Iso. I had seen this rising from the helicopter and it looked impressive. He also claimed another rising from the mountain on the east side. He complained that they did not receive any food or supplies from what we had left for the Ora village people, two years ago as a thank you present.

I gave them some tinned fish and biscuits while we discussed the problem as a diplomatic move. He handed me a letter written in Pidgin language explaining that their ancestors were buried near Mageni so the cave belongs to them. 'Did we pay the Ora people, did we pay two years ago, how much and who did we pay'? My answer was of course 'no'. They wanted anything left to be given to them.

*The author with Alois and the Kol tribe children of Ora village.
Photo: David Nixon.*

I explained about the 1,000 kina which we planned to give to the Ora village and surplus supplies promising that this could be shared. To keep the peace, with Johnny's permission, I gave them the tarpaulin which our Ora men slept under, a water container, cooking pot, 5 kg of rice and 5 tins of fish which they seemed very happy with. The head man did demand that any expedition returning to Mageni must pay them 20,000 kina. I explained that, if this was his demand, no one would come.

Dave, Jean-Paul and William returned to camp around 10am the next day having mapped a further 600 metres of cave passage. After crunching the data in the laptop, Mageni cave now came to an outstanding length of 9 kilometres in length with no end in sight.

On Friday 18th April I was approached by the presenter and one of the film crew to discuss the problem between the two clans; it seemed to me that they were worried that violence could erupt! It was obvious they had little understanding of the problems facing these amazing indigenous groups, living in isolation in the mountains of the Nakanai. I had already settled this problem but to keep the peace I decided to do it again in public rather than in private. I brought the groups together and re-promised the agreement made yesterday; also agreeing to pay every person who carried equipment up the hill to the landing zone 10 kina.

This was greeted with a cheer but sadly was not filmed but again this was the BBC. We packed up the camp and I sorted out 89 loads to be carried up the hill; recording every porter's name, this of course included women and children. That night we camped at the landing zone and the following morning headed off to Palmalmal on seven trips in the helicopter.

It was nice to meet up with Alois, his wife and two children at his house, his work and help was outstanding. We boarded a chartered flight back to Port Moresby and flew back to the UK the following day.

I looked forward to the transmission of the two-part documentary but, although the underground filming was good, the documentary story line was split between two separate areas in Papua New Guinea which was confusing. The film jumped backwards and forwards between the two areas as if they were going on at the same time. The section on walking up through the volcanic debris near Rabual New Britain was about as gripping as watching paint dry.

I have described this filming expedition briefly in one chapter whereas the presenter decided to write a book which was full of inaccurate innuendoes. He claimed that the group arriving from Hobu village was stealing our supplies when, with permission from Johnny, I gave them supplies. This is called diplomacy and being kind. He claimed that Johnny solved the problem! I find it hard to understand that intelligent people fail to understand the difficulties facing other cultures in such isolation. These people have little or nothing, we visit then we disappear.

20

Back to Sarawak

Life would be infinitely happier if we could only be born at the age of eighty and gradually approach eighteen.

Mark Twain

Betty, Jeff, and Racheal always had their annual holiday back in Sarawak to visit all their friends and relatives. When Jeff was 19 years of age, he phoned me from Sarawak to say he was not coming back to the UK. He was old enough to make up his own mind on his future, so I accepted his decision with reluctance. When he was approaching his 21st birthday I thought it would be nice for me to visit him after a 10-year period of my not enjoying the delights of Sarawak. I made enquiries from certain trusted individuals and it became apparent I was not on any blacklist of undesirables. I returned to Sarawak for his 21st birthday and realised how much I had missed the place; it became apparent that now there was no turning back.

Regular visits followed and projects where I could assist on conservation aspects seemed to mount up. I was welcomed back during my visits to the Forest Department in Kuching and worked voluntarily on looking at potential protected areas that would qualify as World Heritage sites. Trans-boundary areas with Brunei Darussalam and Kalimantan, Indonesia all seemed possibilities. Gunung Mulu National Park World Heritage Area could be a part of a trans-boundary area with the protected forest area of Brunei. Lanjak Entimau Wildlife area would qualify for its protection of over 1,000 Orang-utans in the wild bordering Kalimantan. I compiled and submitted project papers along with possible concession management proposals for a few of Sarawak's national parks. Sadly, after years of work nothing came from my proposals.

My stays in Sarawak seemed to expand into many months and I was offered the use of office space by a relative of Betty who owned a construction company. This was a great help as I was able to concentrate on the work in relative peace and quiet.

The tourism sector of Gunung Mulu was now under concession

management and was working well but conservation still came under the Forest Department not the concession company. While I had been in the UK bird nest collecting had mushroomed with many different groups raiding the caves. Unsustainable collecting had drastically reduced the numbers of swiftlets to alarming proportions. After 10 years of illegal collecting, 100 kg of nests had reduced to a few kg and many of the outstanding caves of Mulu had been trashed with rubbish. The Chinese ban on importation of nests thankfully reduced the collecting as the price of nests collapsed so presently 'all is quiet on the western front'. Catching and prosecuting the nesters was almost impossible which was a sad state of affairs for a World Heritage Area of outstanding biodiversity. It is so easy for the nesters to hide. Hunting is still a massive problem in Mulu as the once small population of Penan has swelled to over 500 people out of a total population of around 800. It's easy to see why the area is attractive, a school, clinic, paid work, anywhere you can build a hut and a World Heritage Area next door where they can hunt-is bound to attract the Penan to settle there.

I conducted a major study of the little-known karst area known as the Middle Baram. Over 100 caves are known to exist in this region; many are the habitats of the white nested swiftlet extremely valuable for bird nest soup. All are guarded by the local community to prevent poachers from stealing their nests. This was a long project paper for tourism and conservation. Sadly, the client passed away before any serious work could begin and the area was never protected as a National Park much to my disappointment.

Project planning and design with the construction company started to keep me very busy, especially around micro hydro electrical generation for various Longhouses. I designed a few but it was decided that solar lighting for Longhouses was a cheaper alternative. The big one was a 480-kilowatt mini hydro at Kejin two kilometres from the Baram River where we constructed the generator building and pipeline. I was able to work on this major project for two years. The fun came when we opened the valves to test the water flow and a thrust block moved due to the immense kilonewton (kN) force acting on a bend in the pipeline. I re-designed it with mathematical calculations working out the kilonewton forces via water flow volume and speed, angle of the bend and slope and came up with a figure of tonnage of concrete required and it worked after extensive re-building.

A major tourism and conservation project was the Regional Integrated Highland Development Master Plan for Upper Baram, Miri Division. I was asked to be the Mulu consultant so spent many months working on the project and came up with several tourism projects which could be developed outside the park boundary and operated by a consortium of the local community. Under a cooperative system these projects would receive high visitation, provide employment and a steady income for the community. Unfortunately, no action has been taken yet.

The growing numbers of people settling in Mulu and building houses has reached the stage where the area has been designated as the Sungai Melinau Kampong Mulu with its own Head Man. Proposals are underway for an electrical supply to all houses and a freshwater treatment plant.

The most important conservation project came with my proposal paper on a management plan for Gunung Buda National Park for the Forest Department. Previously I had worked on this project for over six years, but all efforts came to nothing except that the park was finally constituted which I considered as a major achievement. After one year of negotiation a contract was issued and in another two years of work, I produced the management plan. A major part of the plan was a meeting with the local community in an attempt to explain the purpose of the park as regards to conservation and tourism. I felt it important that the local community gained the financial benefits thus helping to ensure long term conservation as this was directly for their benefit. The park had been left as a conservation zone for the 20 years since its conception and little or no enforcement carried out. It was suggested that the park would be in competition with Mulu but in fact this is not the case as all tourist activities would be different to any activities on offer in Mulu. This was the probable reason that no development had taken place.

The caves had been systematically raided for bird nests but, as with Gunung Mulu, the numbers of birds had declined. Frustratingly the allocated funds for tourist development never came to fruition but I am still hoping that one day it will.

With James Cook University, Australia, the book Nakanai Ranges of East New Britain, Papua New Guinea was a project of international importance considering our many years of attempting to have this area conserved with a distinct possibility of World Heritage Nomination. Some progress has been made with a few cultural and natural sites protected. Without doubt the caves are of international significance.

I still have involvement with the International Union Conservation of Nature (IUCN) closely associated with World Heritage natural areas but having Racheal as my globetrotting daughter enabled us to trot around the world looking at World Heritage sites. We were lucky enough to visit Dubrovnik in Croatia, Montenegro and the Mostar Bridge in Bosnia and Herzegovina. Then another year visiting the spectacular World Heritage Cave of Skocjan, in Slovenia and the magnificent Predjama cave with its castle, Lake Bled was a must-see place.

Petra in Jordan, North Africa along with the desert and incredible rock structures in Wadi Rum, the Valley of the Moon was an experience never to be forgotten.

Iceland was a country I had always wanted to see, and we were not disappointed by the dramatic scenery, snow fields, glaciers, and huge waterfalls and not to be outclassed by the whales.

We also had the great opportunity to gaze at the wonders of World Heritage natural areas in Canada. Staying with Daryl and John Donovan whom I had not seen for so many years and staying with the late Chas Yonge. Chas drove us around to see the great sites and we were amazed as he seemed to know the name of every mountain we passed, and he had climbed most of them.

Of course, we did not neglect South East Asia with our time, wandering around in Cambodia at the cultural site at Angkor Wat and elsewhere, also the Killing Fields and the legacy left by Pol Pot and the Khmer Rouge. Another year we had a great trip to Laos where we caved just with a head torch. India was the only country I was not bothered about leaving. Although the Taj Mahal, the palaces and forts are spectacular, the problem is the thousands of people everywhere, air pollution and mountains of rubbish. The country is without doubt over-populated.

One of our best trips had to be a visit to the beautiful country of New Zealand as I had the opportunity to stay with my friend Stan Harding and his wife Ann. It was important as Stan is no longer with us and I had known him from being 16 years of age as we worked together as Radio and TV apprentices. It was with Stan that I started my wanderings, hiking and camping in the Peak District and having many adventures.

I now live most of the time in Miri, Sarawak with my son, his wife Lydia, and my grandson Shayden and where we have a mortgage on a house.

During my 10 years working in the Gunung Mulu National Park my wife's father Usang Ipoi gave us some land where we could build a timber house. It was a nice place with an isolated limestone cliff and my own cave called Gua Usang. Huge fruit trees protected the river banking from erosion. After four years of work we owned a nice five-bedroom wooden house with a couple of bathrooms. All houses need maintenance and especially in the tropics as the downpours and river flooding tend to be severe. The house was on timber posts, above flood level. On returning to Mulu after 10 years away it was sad to see the state of the house as no maintenance had been carried out. A few years ago, a flood damaged the timber posts and the house fell down. All those years of work plus the massive problems and costs of transporting and purchasing all materials seemed to have been for nothing. The house was a write-off.

The problem with land around Mulu is that it requires a house to be built on it. Only then can you claim it as yours.

Our new Mulu house and garden. Photo: David Gill

Betty wanted a house, so I managed to get one built. Only two bedrooms this time but it is raised above ground on concrete posts so the chances of it disappearing in a flood are low. There are no paths or roads, just the Melinau River so I now have a boat; engine and water pump to feed the water tank from the river.

With my brother-in-law Jackson we spend time there now, busy with maintenance work and gardening. It's nice and peaceful listening to the multitude of bird species singing, watching the squirrels on their way home to the cliff in the evening and the cave swiftlets flying back into the cave. I gaze at the river just outside my front door and the surrounding jungle and wave to friends passing in their boats. At night, the bats appear from the cave and hunt around my outside solar flood light for insects while I sit on the veranda and sip a scotch whiskey.

Our Mulu house with Usang Cave to the left. Photo: David Gill

Epilogue

I have been privileged to explore many countries of the globe with the finest country probably being New Zealand, a beautiful country with a brilliant history and people but my fondest memories do belong to Sarawak. Maybe it's the fact that I have spent so many years in this remarkable country or that I married and had two children in Sarawak, or it could be the jungle, the people and the caves.

Coming to terms with conservation and culture is an uphill struggle for the British as we fail to understand the principles governing other people's lives. If you want to understand a culture of a people you need to live with them for a long period of time, not just on a caving expedition for a few months. If they live in a small hut in the jungle, you have to live in a small hut in the jungle. If they use the river as a toilet, you need to use the river as a toilet, wash your cloths in the river, drink the river water, eat the same food, eat wild animals if there is no other choice, live their life and step into their shoes. If they survive in the mountains, or the desert, you have to survive in the mountains or the desert. Only then can you begin to understand their culture.

One thing which is almost impossible to achieve is to convey the basic principles of conservation. All things are interdependent on each other. The cave swiftlet and the bats consume millions of tons of insects or disperse the seeds from fruits helping to keep the balance of the natural world which we are entirely dependent on. The bird nesters kill many thousands of swiftlets via unsustainable methods. Any major interference destroys the balance which is precarious. This is especially true in tropical rainforests. Certain animals reproduce in profusion so it is often necessary to kill and eat, the problem is, which is which. All we can do is to keep trying, failure is not an option.

About the author

David William Gill was born in 1941 in Manchester England. It was regarded as being a hard time during the war with no air raid shelters, bombing, rationing and with his father Frank being called up to serve in the Royal Air force. After the war he moved to Droylsden and in later years became a Radio and TV engineer, beginning as a 15 year old apprentice. He walked and camped in the Peak District as a hobby and took an interest in caves. After getting hooked on the wonders of Derbyshire's caves he became a member of the Eldon Pothole Club based in Buxton joining his first expedition to the Gouffre Berger France in 1967. From gaining experience in the exploration of caves he led his own expeditions to the deepest cave in the world at that time called the Pierre St Martin in the French Pyrenees. This led to 40 different expeditions to many parts of the world including China and Papua New Guinea with the Untamed River expedition to New Britain being one of the most technically difficult and dangerous caves beneath the earth to explore. He ran his own company from his cottage in Chinley Derbyshire as an Electrical Engineer which gave him the freedom to explore and to become involved in writing the Guide books for the Caves of the Peak District and also as a Controller for the Derbyshire Cave Rescue Organisation. From a 25-year-long hobby he was asked at the age of 50 to be the full-time speleologist and development officer for the Gunung Mulu National Park in Sarawak where he worked for a period of 10 years. His major achievements included the establishment of a new protected area, the Gunung Buda National Park that contained 92 kilometres of caves mapped and explored and all the documentation for Gunung Mulu to be nominated as a UNESCO World Heritage Area. After a period back in Manchester he is now living in Sarawak still involved in conservation, world heritage and helping local communities.

Also by David William Gill.

With Paul Deakin. 1975. British Caves and Potholes.
Bradford Barton Ltd. Turo, UK.

With Dr Trevor Ford. 1984. Caves of Derbyshire.
4th edition. Dalesman Books.

With Dr Trevor Ford. 1989. Caves of Derbyshire,
5th edition. Dalesman Books.

1988. The Untamed River Expedition to East New Britain.
Chinley, Derbyshire, UK. ISBN.0.9514004 0 1.

With Dr John Beck. 1991. Caves of the Peak District. Dalesman Books.

1999. The Gunung Mulu National Park Nomination for World Heritage Listing. UNESCO. Sarawak Forest Department, Malaysia.

2011. Untamed Rivers of New Britain Expedition, 2006/2009.
Rainforest Cave and Karst Consultancy, Manchester UK.

2013. A Master Plan, Eco-tourism Development Project Metarae Kejin, Middle Baram Karst, Sarawak. Rainforest, Cave and Karst Consultancy.

With a team of Consultants. 2017 Regional Integrated Highland Development Master Plan for Upper Baram, Miri Division, Sarawak .AJC Planning Consultants Sdn Bhd, Malaysia.

2018. Management Plan for Gunung Buda National Park,
Limbang Division, Sarawak, Malaysia. Forest Department Sarawak, Malaysia. 348 pages.

With a team of Consultants. 2018. The Nakanai Ranges of East New Britain, Papua New Guinea. Cairns. James Cook University, Australia.

2020. Journeys Beneath the Earth. Amazon, KDP print on demand and e-book. 298 pages. The first 25 years.

Printed in Great Britain
by Amazon